T0295154

LIGHTNING IN THE TROPICS: FROM A SOURCE OF FIRE TO A MONITORING SYSTEM OF CLIMATIC CHANGES

CLIMATE CHANGE AND ITS CAUSES, EFFECTS AND PREDICTION SERIES

Climate Change and its Causes, Effects and Prediction Series

Lightning in the Tropics: From a Source of Fire to a Monitoring System of Climatic Changes

Osmar Pinto, Jr.

Nova Science Publishers, Inc.
New York

LIBRARY OF CONGRESS CATALOGING-IN-PUBLICATION DATA

Pinto, Osmar.
  Lightning in the tropics : from a source of fire to a monitoring system of climatic changes / Osmar Pinto, Jr.
      p. cm.
  Includes bibliographical references and index.
  ISBN 978-1-60741-764-4 (hardcover)
  1. Lightning--Tripics--Observations. 2. Tropics--Climate. 3. Global warming. I. Title.
  QC966.P56 2009
  551.56'320913--dc22
                                                                         2009017665

Published by Nova Science Publishers, Inc. ✛ New York

CONTENTS

"Every day there is thunder like I have never heard in my life….and lightning strikes constantly in the mountains which surround the city" written by Marquis of Borba to his family in Lisbon in March 1808, a few weeks after arrival in Rio de Janeiro with Dom João.

PREFACE

Two to four million years ago human beings—known as hominids—lived mainly in the grasslands now known as Kenya, Tanzania and Ethiopia, in the tropical region of the African continent. It is believed that to survive on cold nights they collected the fire set off by a strike of lightning to a stick then carrying the fire with them. When the stick was almost burned out and the fire faded, they would light another stick. Fire was so valuable that, once it was captured, it was carefully tended. If it went out, they had to wait until another strike of lightning set fire to the bush or until they came across other humans who had kept their fire alight. The skilled use of fire—the result of many brainwaves and experiments during thousands of years—is one of the main achievements of human race and the tropical lightning was essential to this achievement. Today we are facing a new challenge—the global warming that humans need to control in order to survive on this planet. And, again, the tropical lightning may play a key role in the process.

Most lightning on Earth occurs in the tropical region, mainly in the central portion of the African Continent, in the Amazon region in South America and in Indonesia. Despite this fact, no book about tropical lightning is currently available. There is also no book about the relation between global warming and tropical lightning. Although this relationship is not very well understood at the present time, it has been considered by the Intergovernmental Panel on Climate Change (IPCC) community as extremely important in the study of global warming. The reason for this is that lightning can contribute through feedback mechanisms to global warming

The purpose of the present book is to review the current knowledge about lightning in the tropical region, trying to compare observations made at different times and places and with different techniques, and indicate how this knowledge can be used to investigate and predict future impacts of global warming on Earth. Most lightning information in the literature is devoted to observations in the temperate region. Therefore the information in the tropical region is relatively scarce and not available in books of general circulation. In this sense, this book contributes to an area where no publication in this form is available; most of the present knowledge is published in specialized lightning-related journals, in general available just to experts in lightning-related fields. This book will be the first to address this subject in a comprehensive way. It will review all available information on lightning in the tropics, organized both in chronological and geographical terms. It will also critically discuss the relation of lightning to global warming, addressing both the effects of global warming on tropical lightning and the role of tropical lightning as a contributing source to global warming.

This book is intended to undergraduate and graduate students, as well as any professional working in fields related to lightning and global warming. These fields include physics, engineering, meteorology, atmospheric physics, space science, mathematics, and any other field where the study of lightning or global warming are relevant. It can be used for teaching, research and as a professional reference.

I have authored three books on lightning, published in 1996, 2002 and 2005 in Portuguese. These three books are primarily concerned with the physics of lightning and lightning observations in Brazil. In this book, I have extended the observations to the whole tropical region and discussed for the first time the relationship between lightning and global warming, an issue which has received increasing interest in the last few years.

This book is divided into 7 chapters. Chapter 1 provides an introduction to the subject of this book, a description of the different types of lightning and general information about the lightning occurrence on Earth. Chapter 2 discusses the basic concepts involved on ground and cloud lightning. Chapters 3, 4 and 5 are primarily concerned with lightning observations in the tropical region by different techniques: satellite, lightning location systems and other techniques, respectively. Chapter 6 considers the relationship between lightning and global warming, addressing different aspects of this relation. Finally, a summary of the status of our knowledge about lightning in the tropics and about our understanding on the relationship between lightning and global warming is given in Chapter 7.

I would like to thank Regina M. Gonzaga dos Santos for reviewing the English usage, my daughter Iara Cardoso de Almeida Pinto for editing all figures and my colleagues of the Atmospheric Electricity Group (ELAT), in particular Iara R.C.A. Pinto, Kleber P. Naccarato and Marcelo M.F. Saba for many helpful discussions on Chapter 4 and Chapter 5, and Earle Williams from the Massachusetts Institute of Technology for helpful discussions on Chapter 6. I would like also to thank Marshall Space Flight Center for kindly providing the OTD and LIS data used in Figures 1.11, 1.12 and 3.2. All other photos in the book (Figures 1.8, 1.9, 2.2, 5.1a, 5.2, 5.6, 5.7, 5.11, 5.12, 5.13 and 6.3) belong to the Brazilian Institute of Space Research (INPE). Finally, I would like to thank INPE for providing support for my research along the last 30 years, which gives me the opportunity to write this book.

Osmar Pinto, Jr.[*]
December 2008

[*] Dr. Osmar Pinto Jr.; Atmospheric Electricity Group, Brazilian Institute of Space Research

Chapter 1

INTRODUCTION

Lightning is perhaps the most beautiful and dangerous phenomenon on Earth. It has fascinated human beings since the first ages, at the same time that it has been deadly. But lightning also had practical applications. In the past, it was fundamental for the survival of human beings on Earth acting as a source of fire. Two to four million years ago human beings—known as hominids—living in the tropical region of the African continent collected the fire set off by a strike of lightning to a stick then carried the fire with them for multiple reasons. This fact can be inferred in many parts from the Bible, where lightning is named "fire from heaven".

More recent practical applications include a forecast tool for intensification of tropical cyclones (Lyons and Keen, 1994; Molinari et al., 1994, 1999; Shao et al., 2005), a complementary tool for supporting studies of lightning related phenomena (Pinto et al., 2004b; Naccarato et al., 2003), an indirect measure of temperature (Williams, 1992) and an indirect measure of concentration of essential elements, such as nitrogen (Bond et al., 2002) and ozone (Ryu and Jenkins, 2005).

At the present time, based on most recent observations, every second on Earth about 50 lightning (or flashes) occur. Most of them occur in the tropical region (Latham and Christian, 1998; Christian et al., 2003). About 30% of all flashes strike the ground somewhere, causing in many cases deaths and damages. It is estimated that worldwide about 20,000 people are killed by lightning every year (Holle and Lopez, 2003), most in the tropical region, in particular in the African continent. Estimates of worldwide losses due to lightning are very uncertain, but from the available information it may amount globally to tens of billions of dollars, 5 billions just in the United States.

In this book, the technical definition of the tropical region, or tropics, which is the geographic region of the Earth centered on the equator and limited in latitude by the two tropics: the tropic of Cancer (23.5° N of latitude) in the northern hemisphere and the tropic of Capricorn (23.5° S of latitude) in the southern hemisphere, is adopted. This definition is in agreement with most of the methods of classifying global climatic zones, which use the tropics of Capricorn and Cancer as bounded lines. Figure 1.1 shows a global map indicating the tropical region and Figure 1.2 shows the different countries in the tropical region for the three main land areas: the American continent, the African continent and the region that includes Indonesia, South Asia and North Australia.

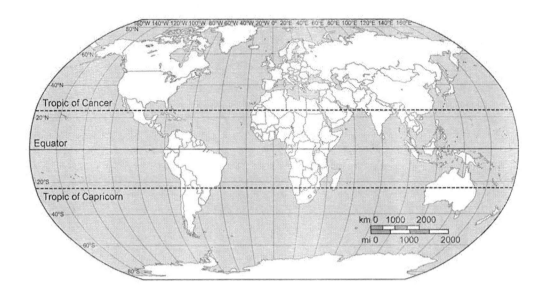

Figure 1.1. World map indicating the tropics of Cancer and Capricorn, which are bounded lines
between which the tropical region is defined.

An analysis of Figure 1.2 shows that the three main land areas in the tropical region have
different features. The South American tropical region is characterized by the largest tropical
country, Brazil, and a large region known as the Amazon basin, which includes the North
region of Brazil and parts of all countries that have borders with this region. The African
tropical continent is a diverse region containing tropical rain forests, alpine zones, and dry
deserts. The region including Indonesia, South Asia and North Australia, in turn, is formed by
many small islands, between two large continental areas. The Indonesian region is sometimes
referred to as the maritime continent.

From the meteorological point of view, the tropical region is characterized at high
altitudes by the trade winds in response to the intense heating by the sun and the convergence
of surface winds from the northern and southern hemispheres along the equator in a region
known as the inter-tropical convergence zone (ITCZ), The convergence causes the air to rise
producing powerful convection. Then, the air spreads out toward mid latitudes, cools off, and
sinks again around 30° latitude flowing back toward the equator, at the same time that the
Earth's rotation causes it to flow westward. The ITCZ migrates north and south of the
geographic equator according to the seasons, showing longitudinal differences due to the
different heat capacity of land and oceans. The tropical region is also affected by many other
phenomena, in particular the El Niño and La Niña phenomena, which are related to variations
in the sea-surface temperature in the equatorial Pacific Ocean. Figure 1.3 shows an
illustration of the trade winds, which are global wind patterns created by the atmospheric
circulation and the Coriolis effect. In response to these meteorological phenomena, tropical
lightning presents large geographical and time variations at many scales. These variations,
both in the present and in the future, are the subject of this book.

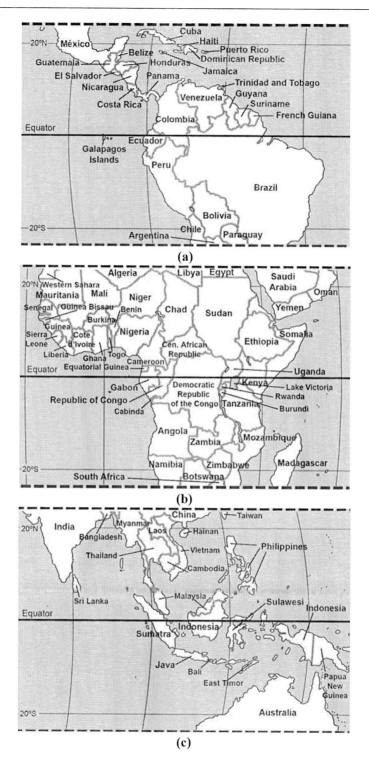

Figure 1.2. Map indicating the different countries in the tropical region for the three main land areas: a) the American continent; (b) the African continent; and (c) the region including Indonesia, South Asia and North Australia. The very small islands are not indicated.

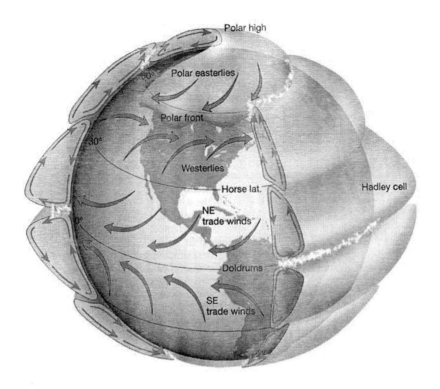

Figure 1.3. Illustration of the trade winds in the tropical region.

1.1. Types of Lightning

Lightning is a very long electrical discharge that occurs in the atmosphere. Most lightning is generated by summer thunderstorms, although they can be produced by volcanoes and dust storms as well (McNutt and Davis, 2000). They occur when the local atmospheric electric field inside a thunderstorm or near a tall object on the ground is able to overcome the atmospheric dielectric rigidity and initiates a corona streamer that leads to lightning initiation.

In a thunderstorm, the electric field is believed to be produced by electrical charges resulting from collisions between soft hail particles and small crystals of ice. Most hail-ice interactions take place at altitudes where the temperature is around -10 °C to -20 °C and in the presence of super cooled water droplets. In a typical thunderstorm, after the charge transfers during the collisions between hail and ice particles, positively charged ice crystals are carried out by updrafts to the top of the thundercloud, to an altitude that varies from 8 km to more than 16 km. They form a main positive charge center, while negatively charged hail remains at the same altitude forming a negative main charge center. As a result, the thundercloud assumes a dipole charge structure consisting of many tens of coulombs of positive charge in its upper portion and an almost equal negative charge at the -10 °C to -20 °C level. In general, a small amount of positive charge is also found near the cloud base, at altitudes where the temperature is near or warmer than freezing, resulting in a tripolar charge structure (Williams, 1989). The origin of this small positive charge center is not clear. Different hypothesis are discussed by Rakov and Uman (2003).

The typical tripolar thundercloud charge structure is illustrated in Figure 1.4. In addition to the three charge centers that form the tripolar structure, screening layers at the top and at the bottom of the thundercloud are included to represent the interaction of the thundercloud charges with the charges in the surrounding atmosphere. Actually, the charge structure of a thundercloud is more complex than shown in Figure 1.4. In general it presents several charge layers almost horizontally displaced (Stolzenburg et al., 1998). Also, the charge structure may vary from storm to storm, sometimes even presenting a structure more similar to an upside-down dipole with the main positive charge center at the bottom and the main negative charge center at the top (Rust et al., 2005). In the tropical region, most thunderclouds occur from around local midday until late evening; at other times, no more than well-separated isolated thunderclouds are likely to develop. In many cases, different thunderclouds interact with each other forming storm systems called Mesoscale Convective Systems (MCSs), which typically have a spatial scale of 100 km or more. They sometimes develop into Mesoscale Convective Complexes (MCCs), although not as frequently as in the temperate regions (Velasco and Fritch, 1987). In the tropical region, fronts are an important feature to trigger thunderstorms mainly in the borders of the regions, and are uncommon at low latitudes near the equator. Throughout the tropics, high ground is an important trigger to thunderstorm formation (Galvin, 2008). Figure 1.5 shows the main high ground areas in the tropics. In the American continent large mountains are located in the west border of the continent (Andes), in the border between Colombia and Venezuela and in southeast Brazil. In the African continent, they occur mainly in the east border. On the other hand in the region that includes Indonesia, South Asia and North Australia, they are distributed throughout almost the whole region.

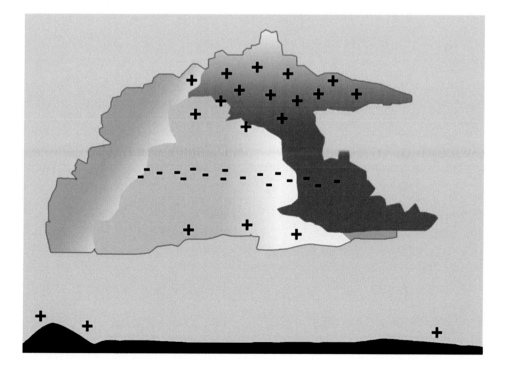

Figure 1.4. Typical tripolar thundercloud charge structure.

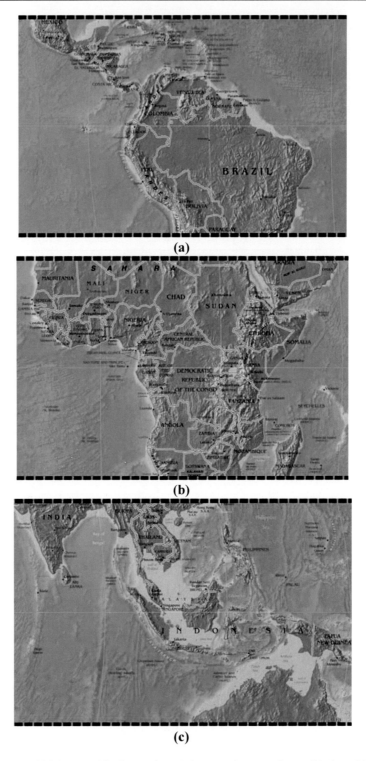

Figure 1.5. Main areas of high ground in the tropics: a) the American continent; (b) the African continent; and (c) the region including Indonesia, South Asia and North Australia. The very small islands are not indicated.

In the tropical region, large storm systems, called tropical storms, form over the ocean depending on the sea-surface temperature. Tropical storms are born over the waters and sometimes disperse to middle latitudes as hurricanes. Lasting from a few hours to as long as a month, they are the largest destructive storms on Earth. However, although lightning data related to tropical storms are limited by the fact that they develop over the ocean, it seems that they are not large lightning-producing storms in a global sense.

All types of lightning can be divided into two categories: those that bridge the gap between the cloud and Earth and contact the ground, called ground flashes, and those that do not, called cloud flashes. Krehbiel et al. (2008) have recently suggested that what defines if a flash will be a ground or a cloud flash is the relative location where the breakdown occurs with respect to the main charge structure of the thundercloud. Figure 1.6 illustrates the two basic types of flashes.

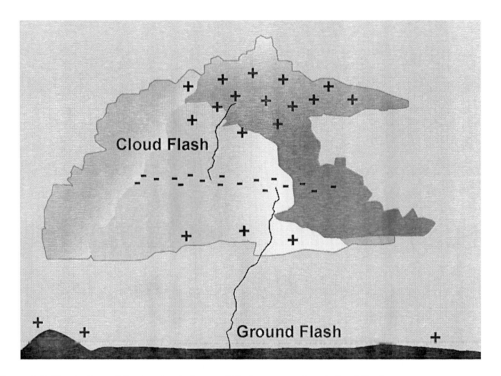

Figure 1.6. Illustration of the two basic types of flashes: ground and cloud flashes.

Ground flashes can be divided into two types as a function of the direction of propagation of the initial process: cloud-to-ground and ground-to-cloud flashes. Cloud flashes, in turn, can be divided into intracloud flashes, which occur totally within a single cloud and are believed to be the most common of all forms of lightning; intercloud flashes, which occur between clouds; and cloud-to-air flashes, which occur between the cloud and the surrounding air. Ground flashes can still be divided according to the polarity of the charge lowered from cloud to ground in the initial process. Thus, there are four types of ground flashes, as illustrated in Figure 1.7. The four types are distinguished from each other by the polarity of charge carried to ground and by the direction of the propagation of the initiation process. They are commonly named cloud-to-ground (or downward) negative and positive flashes and ground-to-cloud (upward) negative and positive flashes. In a few cases, ground flashes present changes in the polarity of the charge lowered to ground during the time evolution and, in these cases, they are called bipolar flashes.

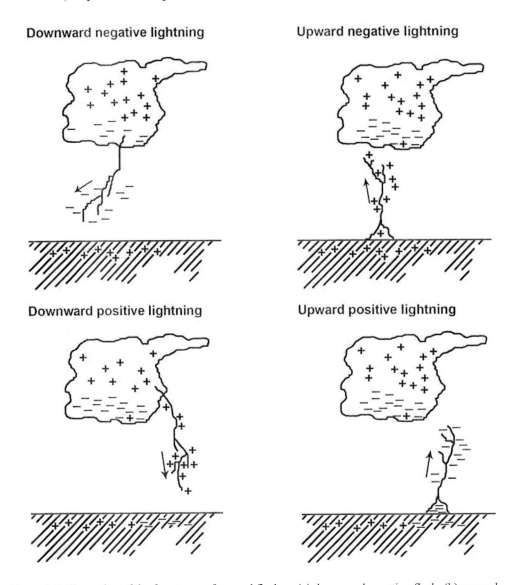

Figure 1.7. Illustration of the four types of ground flashes: (a) downward negative flash, (b) upward negative flash, (c) downward positive flash and (d) upward positive flash.

Cloud flashes can be divided into three different types: intracloud flashes, which are confined within the thunderclouds; intercloud flashes, which occur between thunderclouds; and air discharges, which occur between a thundercloud and clean air.

Figure 1.8 shows the photograph of a cloud flash and Figure 1.9 shows a photograph of a downward ground. Globally, about 90% of ground flashes are downward negative flashes and about 10% downward positive flashes. Upward flashes represent less than 1% of the total number of flashes and typically occur from tall man-made objects higher than 100 m. For objects lower than 100 m or so, upward flashes can still be dominant over downward flashes depending on their effective height. The effective height takes into account the altitude of the object in relation to its vicinity. It appears that negative upward flashes are more frequent than positive upward flashes (Rakov and Uman, 2003).

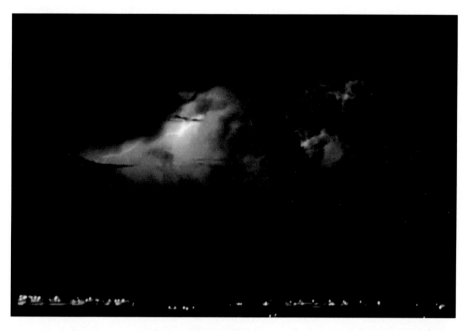

Figure 1.8. Photograph of a cloud flash taken over São José dos Campos, São Paulo, Brazil in 1999.

Figure 1.9. Photograph of a downward ground flash taken over São José dos Campos, São Paulo, Brazil in 2000.

1.2. LIGHTNING OCCURRENCE ON EARTH

The first estimation of the global lightning flash rate, that is, the total number of flashes (both ground and cloud flashes) per second on Earth, was made by Brooks (1925) based on local observations in England made by himself and by Marriot (1908). He estimated a global lightning flash rate of 100 flashes per second, corresponding to a total flash density of about 6 flashes.km^{-2}.year^{-1} over Earth´s entire surface.

However, the first global information available related to the lightning distribution on Earth was obtained only in the 1960s and 1970s, based on the number of days with thunderstorms, or thunderstorm days, per year at different places. This number, also named keraunic level, corresponds to the annual number of days during which thunder is heard at least once at a given location. Figure 1.10 shows a global map of the thunderstorm days as published by the World Meteorological Organization in 1956 (WMO Publication, 1956). The largest thunderstorm day values, above 200 thunderstorm days per year, occur in the following tropical countries: Brazil, Cameroon, Colombia, Java Island, Malaysia, Nigeria, Rwanda, Uganda and Venezuela, among others. From Figure 1.10 it can be seen that most lightning is over the land.

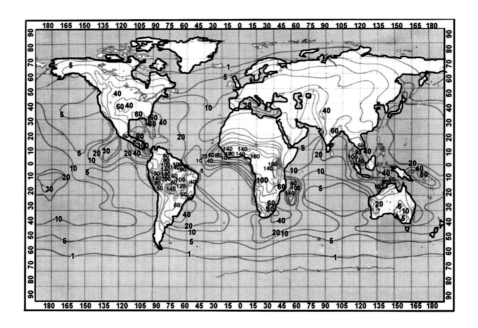

Figure 1.10. A world map of mean annual number of thunderstorm days (adapted from WMO Publication 21, 1956).

More recently, lightning observations by two optical sensors (Optical Transient Detector – OTD and Lightning Imaging Sensor –LIS) on board satellites have shown that about 45 flashes (both ground and cloud flashes) occur around the world every second, about 70% in the tropical region (Figure 1.11). The data in this figure represent a 0.5° x 0.5° composite with a spatial-moving-average operator applied for the 1995-2003 period. They were also corrected by variations in the view time and detection efficiency of the satellite sensors.

Despite most of the tropical region being formed by oceans, satellite observations confirm previous observations of thunderstorm days in which most lightning occurs over the continents (Christian et al., 2003). In support to this fact, there is evidence indicating that thunderstorms over the ocean are less frequent both in space and time (Boccippio and Goodman, 2000) and produce less lightning than those over the continents (Williams et al., 1992; Zipser, 1994; Toracinta and Zipser, 2001; Toracinta et al., 2002). The main physical reason for this pronounced contrast in lightning activity between continents and oceans is generally thought to be a result of the difference in their thermal heating by solar radiation. However, other physical arguments may be important as well, in particular the concentration of aerosols in the air and the variation of the thundercloud base height (Williams et al., 2004). Figure 1.12 shows in more detail the OTD and LIS observations for each continental tropical region: the American continent, the African continent and the region that includes Indonesia, South Asia and North Australia. The largest total lightning activity for each region occurs: in Colombia and Venezuela, in the American continent, Democratic Republic of Congo, Rwanda and Burundi, in the African continent, and Malaysia, in the region including Indonesia, South Asia and North Australia.

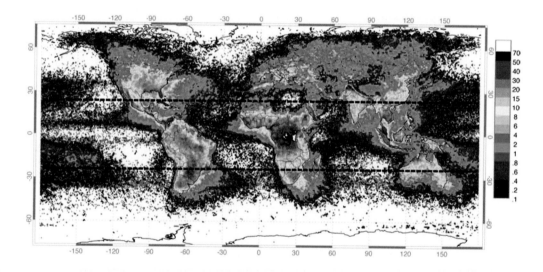

Figure 1.11. Map of the total lightning activity in the world (in flashes.km^{-2}.year^{-1}) combining the observations by OTD sensor from April 1995 to February 2003 and by LIS sensor from January 1998 to February 2003 (courtesy of Marshall Space Flight Center).

Based on recent evidence suggesting that the variation of the cloud to ground lightning ratio is more affected by local thunderstorm characteristics than by latitude (Boccippio et al., 2001; Rakov and Uman, 2003; Pinto et al., 2003a; Chisholm and Cummins, 2006), and that it has values around 2 to 3 in the tropics (see Chapter 4), it is probable that the regions with largest ground lightning activity are the same geographical regions with the largest total lightning activity.

The largest total flash density from the observations in Figure 1.11 occurs in Kamembe, Rwanda, in the African continent, which averages 221 thunderstorm days per year, with 83 flashes.km^{-2}.year^{-1}. The satellite observations in the tropical region will be described in details in Chapter 3.

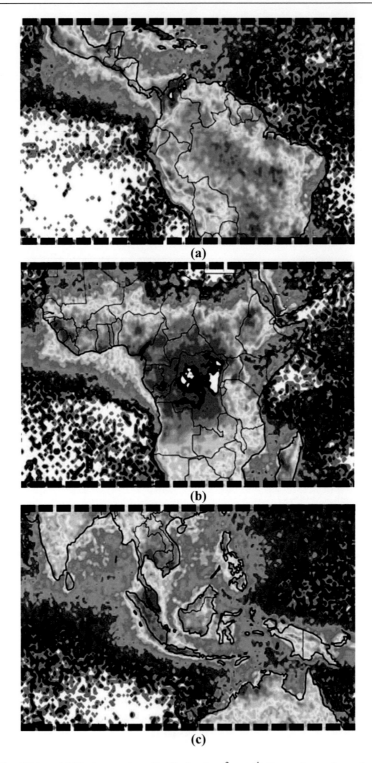

(a)

(b)

(c)

Figure 1.12. The OTD and LIS observations (in flashes.km^{-2}.year^{-1}) for each continental tropical region: (a) the American continent; (b) the African continent; and (c) the region including Indonesia, South Asia and North Australia. The scale is the same as in Figure 1.11 (courtesy of Marshall Space Flight Center).

Chapter 2

BASIC CONCEPTS

Most of the basic concepts about lightning were obtained in three studies conducted in the United States, South Africa and Switzerland in the first half of the twentieth century. The study in the United States was done on the Empire State Building in New York City, a 300 m high structure that is struck by lightning 20 to 25 times per year. About 80% of the flashes striking the structure are upward ground flashes (McEachron, 1939, 1941; Hagenguth and Anderson, 1952). The study pioneered the identification of the main features of the upward ground lightning and was perhaps the first to record lightning current waveforms with oscilloscopes. In contrast, the pioneer work in identifying the main features of downward ground flashes took place in South Africa in the early 1930s, using time-resolved photographic cameras (Schonland, 1956; Malan, 1963). Finally, the study in Switzerland was conducted in two 70 m towers in Monte San Salvatore through resistive shunts to measure directly the lightning current (Berger, 1967, 1977; Berger et al., 1975) and is considered the most comprehensive data on lightning current waveforms ever obtained (Rakov and Uman, 2003).

2.1. GROUND LIGHTNING

The most common ground lightning is the cloud-to-ground or downward negative lightning, which in general is thought to initiate at the bottom of the main negative charge moving in the direction of the small positive charge center beneath it. After some charge neutralization, it continues toward ground where it deposits negative charge. The various physical processes involved in a downward negative lightning are illustrated in Figure 2.1 (Uman, 2008). It initiates with an in-cloud process called the preliminary or initial breakdown, characterized by fast pulses called preliminary breakdown pulses. The conditions inside the cloud involved in the lightning initiation, however, are not well understood (Rakov, 2004). The preliminary breakdown pulses are followed by a less-luminous negative leader called stepped leader, which moves downward in discrete segments (or steps) about 50 m long, with pauses of approximately 50 us between each segment. In Figure 2.1 the steps appear as darkened tips extending downward from the cloud, each step being completed in a microsecond or less. Recent evidence suggests that the negative leader is initiated after a positive leader (Zhang et al., 2008) that moves upward inside the cloud in accordance with the bidirectional theory (Kasemir, 1960; Mazur and Ruhnke, 1993; Mazur, 2002; Williams, 2006). The stepped negative leader branches as it moves downward at an average speed of about 2×10^5 m.s^{-1}, reaching the ground in about 20 ms. In a typical stepped leader, an average current of 0.1 A, superposed by 1000 A rapid pulses, flows along a channel of a few centimeters in diameter transporting on average 5 coulombs of negative charge. Each pulse produces a pulse of visible light, a pulse of electromagnetic energy that peaks in the VHF frequency range and a pulse of X-rays (Dwyer et al., 2005). Figure 2.2 shows an example of a branched stepped leader photographed by a high speed camera in Brazil on its trip to ground.

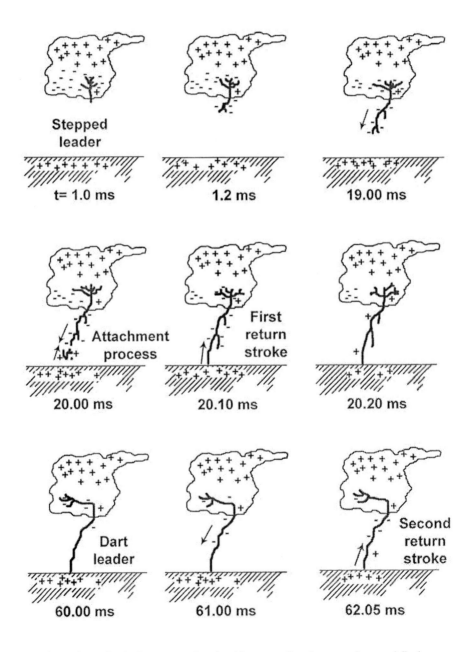

Figure 2.1. The various physical processes involved in a negative downward ground flash.

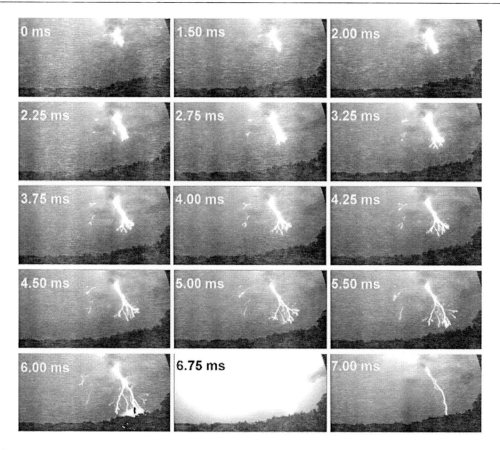

Figure 2.2. Example of a stepped-leader recorded in a sequence of frames obtained by a high-speed camera preceding a return stroke, which is also shown in the last two frames (adapted from Pinto, 2008c).

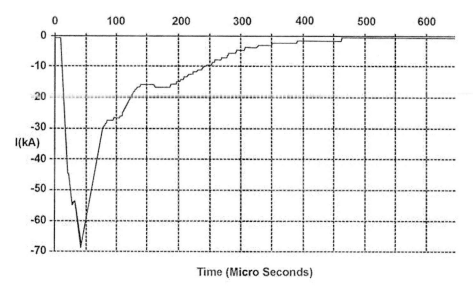

Figure 2.3. Example of a first return stroke for a negative downward flash recorded in the Morro do Cachimbo Station in the Southeast Brazil on March 17, 1992.

When the stepped leader is near the ground, typically tens to a few hundred meters, its presence induces positive charges on ground beneath it, mainly in objects projecting above the Earth's surface. When the induced charge is strong enough to produce an electric field exceeding the breakdown field in virgin air, one or more upward positive connecting leaders from ground initiate and move to join and neutralize the negative charge, in a process called attachment process (see Figure 2.1). When one of the upward leaders contacts a branch of the downward stepped leader, it determines the lightning channel between the cloud and ground. Then, the negative charge near the bottom of the channel moves downward producing a large current pulse called first return stroke (or first stroke). Typically, the stroke has a peak of tens of kA and causes high luminosity and a sound wave called thunder. The channel luminosity and current then propagate continuously upward along the channel (and branches) at a speed typically about one-third to one-half of the speed of light. The return stroke produces very high temperatures along the channel with peaks of the order of 30,000 °C, which ultimately produces the thunder.

After the first return stroke ceases flowing in the channel, the flash may end; in this case it is called a single-stroke flash. However, about 80% of downward negative flashes contain more than one stroke, typically three or four separated on average by 80 ms. The maximum interstroke interval ever reported in the literature was 782 ms (Saba et al., 2006a). Strokes subsequent to the first stroke are called subsequent strokes and are initiated by a continuously propagating leader called dart leader, which moves along the previous channel depositing negative charge (see Figure 2.1) with a velocity of about 10^7 m.s^{-1}. The dart leaders are apparently related to the positive leader which moves upward in the beginning of the process. In general, the dart leader deposits less charge than the stepped leader, resulting in subsequent strokes with lower peak currents when compared to first strokes. In negative downward flashes, subsequent strokes typically have peak currents ranging from 10 kA to 15 kA, while first strokes have typically peak currents from 20 kA to 30 kA. Figures 2.3 and 2.4 show examples of first and subsequent return strokes of a negative downward flash recorded in the Morro do Cachimbo Station in the Southeast Brazil on March 17, 1992, respectively. The first return stroke in this particular flash had a peak current of 70 kA, while the subsequent return stroke had a peak current of 32 kA.

In about half of the times a dart leader, when approaching the ground, changes its path away from the previous leader path giving rise to a dart-stepped leader and a new ground strike point distant from the previous strike point by a few hundred meters to several kilometers (Saraiva et al., 2008b). At present, there is no clear explanation for this feature. Some authors suggest that in some cases after a return stroke residual charge remains in the lightning channel and deviates the new leader from its previous path (Shao et al., 1995; Mazur et al., 1995). The residual charge could be related to the propagation of the positive leader inside the cloud after the return stroke occurs (Mazur, 2002). Because dart-stepped leaders create new paths to ground, a large fraction of downward negative flashes strike ground in more than one point.

In some cases, after a subsequent return stroke an additional current with typical values of tens to hundreds of amperes, called continuing current, flows to ground directly from the cloud. The continuing current represents a relatively steady charge flow between the main negative charge center and ground. Occasionally, this may also occur after a first return stroke. About 35% of all downward negative flashes contain at least one continuing current exceeding 10 ms (Shindo and Uman, 1989; Saba et al., 2006a) and 10% contain at least one continuing current exceeding 40 ms usually referred to as a long continuing current (Saraiva et al., 2008b). Maximum values for the continuing current duration range from 500 to 600 ms (Ogawa, 1995; Mazur et al., 1995; Saba et al., 2006a). In general, during the continuing current intervals slow enhancements in current and luminosity called M-components occur.

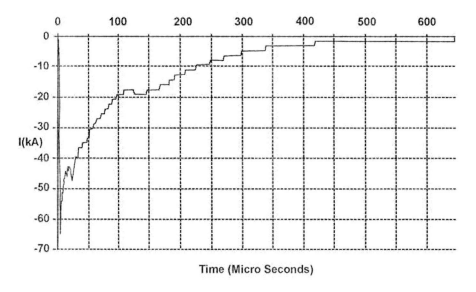

Figure 2.4 – Example of a subsequent return stroke for a negative downward flash recorded in the Morro do Cachimbo Station in the Southeast Brazil on March 17, 1992.

As a whole, the various processes in a negative downward flash emit electromagnetic radiation with a peak in the radio-frequency spectrum at 5 to 10 kHz when observed at distances beyond approximately 50 km. At higher frequencies, the spectral amplitude is approximately inversely proportional to the frequency up to about 10 MHz and, after this frequency, inversely proportional to the square of frequency up to 10 GHz (Cianos et al., 1973).

Thus far in this section the main characteristics of downward negative flashes, which represent most of ground flashes have been discussed. Positive downward ground flashes, that is, flashes that transfer positive charge to ground, are the second most common type of ground flashes. They have characteristics that differ somewhat from those of negative downward flashes. Positive downward flashes are initiated by less stepped leaders, probably due to the differences in the propagation of negative and positive leaders (Williams, 2006), although the leader velocity is not much different from that of negative stepped leaders (Saba et al., 2008b). Positive downward flashes have recently attracted considerable attention for many reasons (Rakov, 2007). Some of them are: they are associated with the highest recorded lightning currents, near 300 kA, and the largest charge transfers to ground, hundreds of coulombs or even more (Goto and Narita, 1995); they have been found to be preferentially related to luminous phenomena in the middle and upper atmosphere known as sprites (Lyons et al., 1998b); and they are predominant in thunderstorms for some particular meteorological conditions (Pinto et al., 1998). For large regions, however, the mean peak current for positive downward flashes is apparently slightly more intense than for negative downward flashes. In addition, positive downward flashes are normally single stroke flashes and tend to be followed by continuing currents longer and more intense than negative downward flashes.

Differently from downward ground flashes, upward ground flashes are initiated by upward leaders not followed by a stroke similar to the first return stroke of downward ground flashes. Rather, the upward leaders are followed by long continuing currents, called initial continuous currents, typically lasting some hundreds of milliseconds, superposed by fast pulses similar to subsequent strokes called initial stage current pulses. After this initial current has ceased to flow, in about 25 to 50% of the cases sequences of downward dart leaders and return strokes very similar to those of downward negative ground flashes occur (Rakov, 2003). Continuing currents and M-components similar to downward flashes are also observed. Berger and Vogelsanger (1969) suggested that the initiation of an upward ground flash is preceded by a cloud flash. Upward ground flashes are very similar to flashes artificially initiated with the rocket-and-wire technique.

Ground flashes that transfer both negative and positive charges are referred to as bipolar lightning flashes. Most bipolar flashes were identified in direct current measurements on tall grounded objects. Bipolar ground flashes with lightning current waveforms exhibiting polarity reversals were first reported by McEachron (1939, 1941) in the Empire State Building observations. Later on they were observed in many other countries (Gorin et al., 1977; Berger, 1978; Gorin and Shkilev, 1984; Heidler et al., 2000).

There are basically three types of bipolar lightning flashes: those with polarity reversal during the initial continuous current, typically in upward or triggered flashes; those with different polarities of the initial continuous current and of the following return stroke; and finally those that involve return strokes of opposite polarity. In some events bipolar flashes may belong to more than one of the above categories. A more detailed discussion about bipolar flashes can be found in Rakov and Uman (2003) and Rakov (2007).

2.2. CLOUD LIGHTNING

The term cloud flashes is used to denote three different types of flashes: intracloud flashes, which are confined within the thunderclouds; intercloud flashes, which occur between thunderclouds; and air discharges, which occur between a thundercloud and clear air. About 75% of the flashes on Earth are cloud flashes. Most cloud flashes are believed to be of the intracloud type, although no reliable statistical data are available in the literature to confirm that this is the case (Rakov and Uman, 2003).

Cloud flashes are less studied than ground flashes for many reasons, including the difficulty to obtain photographic records of the lightning channel or direct current measurements, and the fact that they produce less damage than ground flashes. More recently VHF lightning location systems have made it possible to obtain time-resolved images of cloud flashes (Rison et al., 1999, 2005; Thomas et al., 2000) increasing considerably our knowledge about them.

Cloud flashes are believed to have two distinct stages: an early active stage and a late final stage. The early stage has a typical duration of tens to a few hundreds of milliseconds, and initiates with a negative leader similar to the stepped leader of a negative downward ground flash moving upward and a positive leader moving horizontally in the region of the main negative charge center, forming a bidirectional leader. The beginning of this stage is marked by larger bipolar fast pulses, some of them even larger than preliminary breakdown pulses in ground flashes called compact intracloud discharges (Smith et al., 1999). The transition from the early to the late stage is thought to be associated with the loss of connection between the negative and positive leaders. Cloud flashes usually exhibit more variability than ground flashes, since the latter involve a relatively well-conducting ground "electrode" (Rakov and Uman, 2003). An extensive description of cloud flashes can be obtained in Rakov and Uman (2003).

Chapter 3

SATELLITE OBSERVATIONS

3.1. OPTICAL SENSORS

With the advent of Earth-orbiting satellites, it is now possible to measure lightning from the space by detecting the optical or the radio signals emitted to space by both ground and cloud flashes. The first optical sensor on board a satellite to measure lightning was apparently launched on the Orbiting Solar Observatory satellite - OSO-2 (Rakov and Uman, 2003). However, the sensor recorded lightning only during night-time and, hence, a small fraction of a few percents was detected (Christian et al., 1989).

The first comprehensive satellite study of the global lightning distribution used a high-resolution scanner that produced visible photographs at an altitude of 830 km in the Defense Meteorological Satellite Program (DMSP) satellite (Orville and Henderson, 1986). These observations were followed by many others, which are summarized by Christian et al. (1989).

It was only in the 1990s with the Optical Transient Detector (OTD) and Lightning Imaging Sensor (LIS) launched by NASA in low Earth orbits that lightning was recorded from space during daytime, although like previous satellite-based optical sensors they cannot distinguish between ground and cloud flashes. Both sensors use narrow band optical filters to select an oxygen triplet line generated by lightning centered at 777.4 nm. The OTD was launched on the Microlab-1 (later renamed OV-1) satellite in April 1995 into a 735 km altitude 70° orbit, orbiting the Earth in 100 minutes (Christian et al., 2003). The sensor operated for five years and stopped sending data in April, 2000. It had a 100° field of view, observing a 1300 km x 1300 km region, about 1/300 of the Earth's surface at any given instant, with a nominal spatial resolution of about 10 km and nominal time resolution of 2 ms. Because of its orbit, the OTD observes a given location for just a few minutes per day. A comparative study with the National Lightning Detection Network data in the United States determined that OTD detection efficiency from ground flashes occurring inside its view field was about 45% to 70%, being slightly higher for cloud flashes. The spatial errors are, on average, about 20 km to 40 km, and the temporal errors are less than 100 ms (Boccippio et al., 2000). OTD data must be averaged over 55 days to minimize the diurnal lightning cycle bias.

The second sensor, LIS, was launched aboard the Tropical Rainfall Measuring Mission (TRMM) observatory in November 1999 into a 350 km altitude 35° orbit. It views a 600 km x 600 km and observes a given location for about 3 minutes per day. Due to its inclination, LIS can observe lightning only in a latitude belt between 35° S and 35° N. The average LIS detection efficiency was estimated to be about 80-90%, with a small variation from daytime to night-time.

Since OTD and LIS sensors can gather lightning data under daytime conditions as well as at night, they provide a much higher detection efficiency and spatial resolution than was attained by earlier satellite lightning sensors. The instruments record the time of the lightning events, measure their radiant energy, and determine the location of lightning events within its field-of-view. More details about the sensors can be found in Boccippio and Goodman (2000) and Christian et al. (2003).

Up to date, all satellite optical sensors have recorded only a small fraction of the lightning flashes because they are in a relatively low orbit and hence spend a short time (tens to a few hundreds of seconds) over any given location. During the next decade NASA is planning to put a lightning sensor called Geostationary Lightning Mapper (Christian et al., 1989; Christian, 2007) in a geostationary orbit. The sensor is a charge-coupled device (CCD) designed to detect about 90% of both ground and cloud flashes and will provide coverage over much of North America, Central America and part of South America, covering part of the tropical region.

Other satellites have measured lightning even though they were launched for various other reasons. One of them is the Fast On-Orbit Recording of Transient Events (FORTE) satellite, launched in 1997, containing sensors both in optical and radio frequency ranges (Jacobson et al., 1999, 2000; Suszcynsky et al., 2000; Boeck et al., 2004). Kirkland et al. (2001) have shown, however, that the FORTE detection efficiency for ground flashes is very low (7%). In addition, although in the case of radio-signals the detection efficiency is almost constant during the day, a significant fraction of the data may be contaminated by man-made sources on the ground (Kotaki and Katoh, 1983). The Imager of Sprites and Upper Atmospheric Lightning (ISUAL) experiment onboard the FORMOSAT-2 satellite (Chen et al., 2008) uses three different optical sensors to record transient luminous events related to lightning, allowing the inference of some lightning characteristics related to these events.

3.2. OBSERVATIONS

The first observations of lightning from space by Orville and Henderson (1986) are shown in Figure 3.1. They have studied the global distribution of about 32,000 flashes at midnight between 60° S and 60° N for about one year. Figure 3.1 shows the geographical distribution of midnight lightning throughout the year, in the northern hemisphere summer and in the southern hemisphere summer. They confirm that lightning occurs mainly over the continents, previously inferred from thunderstorm day data, and that the lightning activity during the northern hemisphere summer is slightly higher than during the southern hemisphere summer.

Later on, Orville and Spencer (1979) using data from two DMSP satellites estimated a global flash rate of 123 flashes.s^{-1} at dusk and 96 flashes.s^{-1} at midnight. They also confirmed that there are more lightning during the northern hemisphere summer than during the southern hemisphere summer. Other lightning observations by satellite were carried out by Turman and Edgar (1982) on DMSP and Kotaki and Katoh (1983) on ISS-b.

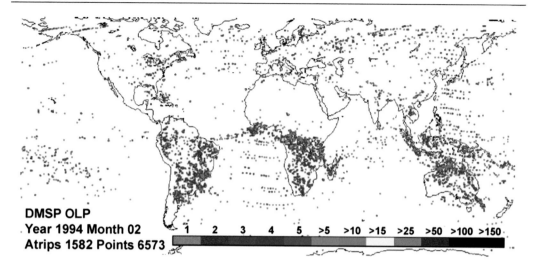

Figure 3.1. Geographical distribution of midnight lightning: (a) throughout the year; (b) in the northern hemisphere summer; and (c) in the southern hemisphere summer observed by the DMSP satellite (adapted from Orville and Henderson, 1986).

In the 1990s, satellite observations by OTD and LIS indicate that most of the lightning on Earth occurs in the tropical region (Latham and Christian, 1998; Christian et al., 2003). About 70% of the flashes occur in the tropics and 40% between 10° S and 10° N. Figure 3.2 shows the total flash density observed by the OTD sensor from 1995 to 2000 (a more updated map including LIS observations is shown in Chapter 1). Flash rates are based on a 0.5° x 0.5° composing grid smoothed with a 2.5° spatial moving average operator. From the data in Figure 3.2 a global total flash rate of about 45 flashes.s^{-1} was estimated, with an uncertainty of 5 flashes.s^{-1}, corresponding to more than 4 million flashes per day and more than 1.5 billion flashes per year.

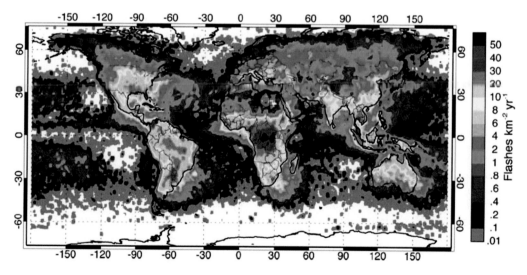

Figure 3.2. A global map of total lightning flash density in flashes.km^{-2}.year^{-1} based on the OTD observations from 1995 to 2000 (courtesy of Marshall Space Flight Center).

In spite of the major part of the tropical region being oceans, total lightning observations by satellite suggest that most tropical lightning occurs over the continents, about 10 times more than over the oceans (Christian et al., 2003). This difference can be explained considering that thunderstorms over the ocean are less frequent both in space and time (Boccippio and Goodman, 2000) and produce less (a factor of two) lightning than those over the continents (Williams et al., 1992; Zipser, 1994; Toracinta and Zipser, 2001; Boccippio and Goodman, 2000; Toracinta et al., 2002). The physical reason for this pronounced contrast in lightning activity between continents and oceans is generally thought to be a result of the difference in their thermal heating by solar radiation, which is, ultimately, responsible for the larger updraft velocities in continental thunderstorms than in oceanic thunderstorms (Price and Rind, 1992; Williams and Stanfill, 2002; Williams et al., 2004). Williams et al. (2004) have addressed the possible role of the order-of-difference in the concentration of cloud condensation nuclei (CCN) between the continents and oceans on this pronounced contrast, using islands as miniature continents. They concluded that this role, termed aerosol hypothesis, is less important than the thermal hypothesis.

Table 3.1. The approximate locations having the greatest mean annual flash density in the tropical region based on OTD data.

Rank	Place Name	Flash Density (flashes.km^{-2}.year^{-1})	Latitude (°N)	Longitude (°E)
1	Kamembe, Rwanda	83	-1.25	27.75
2	Boende, Dem. Rep. Congo	66	0.25	20.75
3	Lusambo, Dem. Rep. Congo	52	-4.75	24.25
4	Kananga, Dem. Rep. Congo	50	-5.75	18.75
5	Kuala Lumpur, Malaysia	48	3.25	101.75
6	Calabar, Nigeria	47	5.25	9.25
7	Franceville, Gabon	47	-2.25	14.25
8	Ocana, Colombia	40	8.25	-74.75
9	Conception, Paraguay	37	-23.25	-57.25
10	Aranyaprathet, Thailand	36	13.75	102.25
11	Miandrivaso, Madagascar	36	-19.25	45.75
12	Manfe, Cameroon	36	5.7	8.3
13	Kindia, Guinea	35	10.75	-12.75
14	Murree, Pakistan	33	33.75	73.25
15	Fitzroy Crossing, Australia	33	-16.25	126.25
16	Bahar Dar, Ethiopia	33	12.25	36.75
17	Campo Grande, Brazil	33	-21.25	-53.75
18	Loc Ninh, Vietnam	32	11.75	106.75
19	Porto Nacional, Brazil	32	-10.25	-47.75
20	Ouanda Djalle, Central African Republic	32	8.25	-10.25
21	Macenta, Guinea	31	8.25	-10.25

Total lightning observations by satellite also reveal that there is more lightning in the northern hemisphere than in the southern, because the first has a larger land area. This asymmetry, however, is not seen in the tropical lightning activity. On the diurnal basis, in turn, continental lightning peaks in the afternoon around 16:00-17:00 local time, with a minimum activity in the early morning hours. Ocean lightning, however, is equally distributed during the day in consequence of its almost constant temperature throughout the day.

Among the three continental regions, the largest lightning activity occurs in the African continent, followed by the American continent and Indonesia. The peak in the lightning activity occurs in Kamembe, Rwanda, over the equatorial Congo basin in the African continent, with 83 flashes.km^{-2}.$year^{-1}$. Table 3.1 shows the locations (within the spatial resolution of 0.5° x 0.5°) with average annual flash densities higher than 30 flashes.km^{-2}.$year^{-1}$ within the tropical continents based on the OTD data. The largest value observed by OTD outside the tropics was in Posadas, Argentine of 43 flashes.km^{-2}.$year^{-1}$.

The OTD data also allowed the verification that the largest flash density occurs in the Congo basin at any time throughout the year and the largest intraseasonal variations observed in the tropics occur in northwest Australia and in the north central Madagascar. In addition, the global tropical flash rate over land shows an apparent semiannual peak, while the global oceanic flash rate is almost constant during the year. The semiannual peak has been confirmed by lightning observations from lightning location systems (Williams et al., 2002; Abdullah and Yahaya, 2008; Torres et al., 2008) and thunderstorm days (Kandalgaonkar et al., 2005) near the equator, although it is not present in some observations (Molinié and Pontikis, 1995). In contrast, near the edge of the tropics the semiannual peak is sometimes observed (Pinto et al., 2003b) sometimes not (Kuleshov et al., 2006; Pinto et al., 2007).

Williams and Stanfill (2002), Williams and Sátori (2004) and Williams (2005) have addressed in details the reason for the peak in the global lightning activity in the African continent. The ranking of the different tropical regions in lightning clearly indicates that Africa is the most continental chimney; Indonesia, South Asia and North Australia are the most maritime (less lightning) "continent", with the American continent as intermediate. The American continent is also strongly maritime during the westerly wind regime of the wet season (Williams et al., 2002). It is, for this reason, dubbed the "green ocean". The reasons for the African continent lightning dominance are apparently related to two if its characteristics: its drier surface allows for the development of a deeper reservoir of unstable air over the course of the diurnal cycle and a higher cloud base height; and its more polluted boundary layer keeps the cloud droplets small and allows more liquid water to enter the mixed phase region where it can invigorate the ice-based electrification process (Williams, 2005). Price (2008) has also observed another aspect of the African continent: its higher elevation above the sea level. Zipser et al. (2006) found from the analysis of the LIS data that the thunderstorms in the African continent are more intense than in the Amazon region and Indonesia.

Naccarato et al. (2008) used ten years of LIS data (1998-2007) to make a ground lightning map for Brazil with a 25 km x 25 km resolution, after correction of LIS data due to the diurnal variation of the detection efficiency and the differential sampling over the tropics compared to the equator. The final values are computed considering a cloud to ground flash ratio of 1.5. Their map is shown in Figure 3.3. Similar maps have been generated for other countries in the tropical region (Kuleshov et al., 2006).

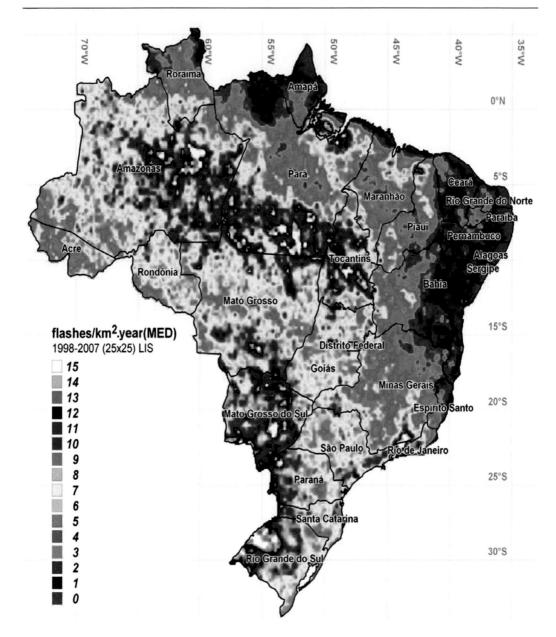

Figure 3.3. Map of ground flash rate over Brazil based on 10 years of LIS data (1998-2007), after correction of LIS data due to the diurnal variation of the detection efficiency and the differential sampling over the tropics compared to the equator. The final values are computed considering a cloud to ground flash ratio of 1.5 (adapted from Naccarato et al., 2008).

Recently, Chen et al. (2008) have inferred from the observations of transient luminous events related to lightning by the Imager of Sprites and Upper Atmospheric Lightning (ISUAL) experiment, onboard the FORMOSAT-2 satellite, that the relative occurrence of flashes with peak currents larger than 80 kA is ten times higher over the ocean than over land.

Chapter 4

OBSERVATIONS BY LIGHTNING LOCATION SYSTEMS

Ground lightning observations by Lightning location systems (LLS) are restricted to a small part of the tropical region. In particular, no observations exist over the equatorial Congo basin in the tropical region of the African continent, where the peak of the total lightning activity occurs (Christian et al., 2003). Also, no cloud lightning observations by LLS exist in the tropical region.

4.1. LIGHTNING LOCATION SYSTEM TECHNIQUES

Lightning location systems using electromagnetic radio-frequency at different frequency ranges from a few kilohertz to a few hundred megahertz (VLF to VHF) have been in operation over many decades to detect and locate all types of flashes (Rakov and Uman, 2003; Diendorfer, 2008a). In general, they have been used in many applications by power utilities, weather services, aviation, geophysical research and others.

In the VLF range (3 kHz to 30 kHz) LLS can cover large areas (Lay et al., 2004, Rodger et al., 2004, 2005; Jacobson et al., 2006) since at this frequency range the signal can circle the globe without too much attenuation by repeatedly bouncing off the conducting ionosphere and Earth. However, they obtain only poor information about the flashes, not discriminating ground and cloud lightning. Although most observation is believed to be associated with ground flashes, no polarity and peak current are estimated for the flashes. Also, the location accuracy is limited and only about 1 to 10% of the flashes are detected. These LLS will be discussed separately in Chapter 5.

In the VHF range (30 to 300 MHz) LLS are not primarily intended to locate ground flash striking points, but rather to image the whole lightning channel, both inside and outside the thunderstorm, or to detect cloud flashes. They are in operation in many countries, including in the temperate regions of the United States, France, Spain, Poland and Japan. In some cases the VHF LLS operate in both VLF/LF and VHF and are designed to detect all types of flashes. In the tropical region, however, no VHF LLS have operated until now.

The most accurate LLS for determining the ground lightning striking point and estimating the peak current of the flash operate in the VLF/LF range (3 kHz to 300 kHz). In this range the LLS are designed to detect mainly ground flashes and are in operation in many countries, including in the temperate region of the United States (Cummins et al., 1998a, b; Zajac and Rutledge, 2001; Orville and Huffines, 2001; Orville et al., 2002), Austria (Diendorfer et al., 1998; Schulz et al., 2005), China (Chen et al., 2002, 2004), Spain (Soriano et al., 2001, 2005), Canada (Burrows et al., 2002), Italy (Bernardi et al., 2002), Japan (Shindo and Yokoyama, 1998; Suda et al., 2002), South Africa (Ndlovu and Evert, 2004; Evert and Schulze, 2005; Bhikha et al., 2006; Gill, 2008), Germany (Gorbatenko et al., 2008), Poland (Loboda et al., 2008) and in the tropical regions of Brazil (Pinto, 2003, 2005, 2008a, b; Pinto et al., 2004a, 2007, 2008), Colombia (Torres et al., 2001; Younes et al., 2003, 2004), Venezuela (Raizman et al., 2004; Tarazona et al., 2006), Java Island (Hidayat et al., 1996; Hidayat and Ishii, 1998, 1999; Berger and Zoro, 2004), South China (Chen et al., 2002, 2004), Papua New Guinea (Orville et al., 1997), North Australia (Peterson and Rutledge, 1992; Sharp, 1999; Kilinc and Beringer, 2007) and Malaysia (Abdullah and Yahaya, 2008).

LLS operating in the VLF/LF range consist basically of several sensors, which determine the angle to the ground lightning stroke at the sensor location and/or the time of the stroke event, and a processing unit, which calculates stroke characteristics such as the strike point location and time, and estimate the stroke peak current, rise-time and pulse width.

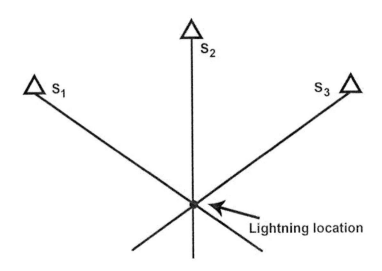

Figure 4.1. Lightning location using three MDF sensors.

The first VLF/LF lightning sensor developed, the magnetic direction finder (MDF), was composed of two vertically orthogonal loops of wire, oriented NS and EW, used to determine the direction to a vertical current source, correspondent to the lower part of the stroke channel. The output voltages from the loops are proportional to the components of the magnetic field in the respective directions. These sensors have been extensively used since 1920s (Horner, 1954) at a narrow VLF band as the primary means of identifying and mapping thunderstorms. In the 1970s, gated wideband direction finders were developed by sampling the magnetic field components of the return stroke field at the initial peak, corresponding to the initial peak of the stroke current (Krider et al., 1976).This was done to overcome the large position errors inherent in the narrowband direction finder approach for locating lightning. In the beginning of 1980s the MDF systems incorporated an electric field sensor that allowed the detection of polarity of the flash, producing the first LLS observations of positive ground flashes (Rust et al., 1981). At least three MDF sensors are necessary to locate a stroke without ambiguity, as indicated in Figure 4.1. The first commercial MDF system produced by the Lightning Location and Protection, Inc (LLP) was introduced in 1970s. The first LLP system operated in the United States at the end of 1970s (Krider et al., 1980).

In fact, the intersections of each pair of sensors yield a different location due to random and systematic angular errors. The latter is generally caused by electrically conducting objects near the sensor. An optimal estimate of the actual location of the striking point, also called most probable location, can be found using a least-square minimization error technique (Koshak et al., 2004).

In the 1960s, a new sensor called time-of-arrival (TOA) sensor (Lewis et al., 1960) was developed. This sensor uses a gated wideband electric field sensor to measure the time at which a pre-determined portion of the field, in general the peak electric field, arrives at the sensor. The time difference between two sensors defines a hyperbola passing between the two sensors which corresponds to the locus of all points where the source (stroke) can be located. To have an unambiguous location for a stroke, at least four TOA sensors must detect the stroke. The lightning location can also be obtained from the intersection of circles generated from the sensor locations, as indicated in Figure 4.2. The location error in the TOA technique is dependent on the time synchronization among the different sensors. Like the MDF systems, TOA systems use the same least-square minimization error technique to obtain an optimal location. The first commercial TOA system, called Lightning Positioning and Tracking System (LPATS), was introduced in the 1980s by the Atlantic Scientific Corporation, which later became Atmospheric Research Systems, Inc. - ARSI (Lyons et al., 1989). The system measures the electric field component of the stroke electromagnetic field at similar frequencies as the wideband magnetic direction finder. The first LPATS LLS used satellite signals for time synchronization and did not employ any method of discrimination against cloud discharges or non-lightning sources, implying a low reliability data. Later on, the TOA systems began to use the GPS timing systems, which kept the error margin very small.

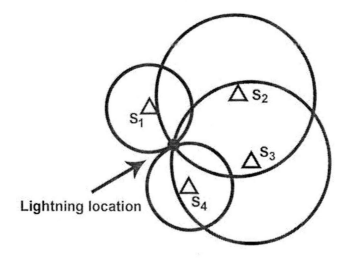

Figure 4.2. Lightning location using four TOA sensors.

In the early 1990s, a new sensor combining both techniques (MDF and TOA) was developed by a company named Global Atmospherics, Inc. (GAI), now, Vaisala Inc., obtaining a better performance compared to each individual technique. The algorithm for lightning location also combines information provided by both techniques. When using these sensors, only two of them are required for an unambiguous location (Orville, 1995). The first commercial sensor was called Advanced Lightning Direction Finder (ALDF). Later on, a new sensor called Improved Accuracy from Combined Technology (IMPACT) was developed (Cummins et al., 1998a, b) and more recently a new version of the IMPACT sensor, LS7000, was developed.

The LLS performance, however, depends on many other parameters in addition to the type of sensors, such as the sensor geometry, gain and trigger threshold. It also depends on the algorithm for processing the data, which includes the waveform acceptance criteria for ground flashes, the algorithms for flash grouping and peak current estimation, among others. In general, the performance of LLS are evaluated based on three parameters: the false alarm rate, that is intimately related to the contamination by cloud flashes, interpreted erroneously as ground flashes, the location accuracy and the detection efficiency, that is, the percentage of flashes (or strokes) occurring that are detected. The cloud flash contamination of ground LLS data depends basically on the criteria used to discriminate them, which in turn are sensitive to other parameters including sensor type, etc. For most recent LLS the contamination is limited mostly to positive downward ground flashes observations with peak currents below 10-20 kA, while in the past networks it usually extended to large peak currents and both lightning polarities. The location accuracy is normally characterized as an ellipse error centered in the optimized stroke location and circumscribing a region, within which there is a definite probability that the stroke occurred. The detection efficiency is discussed in section 4.3.

4.2. LLS IN THE TROPICAL REGION

The first LLS in the tropical region (hereafter referred to as Brazil Southeast-1) was installed in southeastern Brazil in 1988, covering the region from 16° S to 22° S and from 42° W to 48° W (Pinto et al., 1996, 1999a, b). It operated initially with four and later 6 LPATS III sensors. In 1996 it was upgraded to include IMPACT sensors and also expanded to cover a larger region. The number of sensors increased gradually to 24 sensors in mid 2005, when it became the largest LLS in the tropical region (hereafter referred to as Brazil Southeast-2). The Brazil Southeast-2 network was the first LLS in the tropical region to be validated in terms of location accuracy and detection efficiency (Ballarotti et al., 2006) and calibrated in terms of the gain of the sensors (Naccarato, 2005). More details can be found in Pinto (2003, 2005, 2008a), Pinto and Pinto (2003), Pinto et al. (2003a, b, 2006a, 2007a, b), and Naccarato et al. (2003).

In 1999, another LLS was installed in the North region of Brazil (hereafter referred to as Brazil North), with the main goal of providing true ground data for the Lightning Imaging Sensor (LIS). It operated up to 2005 when it was integrated to the LLS in southeastern Brazil (Pinto and Pinto, 2003; Pinto et al., 2003a; Blakeslee et al., 2003; Fernandes, 2005).

With the integration and the expansion of the network to the South, Center and North regions of Brazil, a new LLS called Brazilian Lightning Detection network (BrasilDat) (hereafter referred to as Brazil BrasilDat) began its operation in 2007. The Brasildat LLS has 35 sensors in the tropical region (12 IMPACT and 23 LPATS) and a total of 46 sensors covering approximately two thirds of the country. It is the largest LLS in the tropical region (Pinto et al., 2006b, 2007a, b, 2008a). Figure 4.3 shows the sensor configuration of the BrasilDat in the beginning of 2008.

Figure 4.3. The location of the BrasilDat sensors in the beginning of 2008.

The LLS in tropical Australia was installed in 1998 with 16 LPATS-III sensors (Peterson and Rutledge, 1992; Sharp, 1999; Kilinc and Beringer, 2007). The LLS in Colombia was installed in 1997. It was composed of 6 LPATS-III sensors and operated until 2003 (Torres et al., 2001; Younes et al., 2003, 2004). The LLS in Venezuela began its operation in 2000 with 12 IMPACT sensors, covering the whole country (Raizman et al., 2004; Tarazona et al., 2006). The LLS in Java Island began its operation in 1994, operating with four ALDF sensors (hereafter referred to as Java Island-1). In 1995, it was changed to 8 LPATS sensors (hereafter referred to as Java Island-2) and in 2004 it was upgraded with some IMPACT sensors; nevertheless, no data is available after 2004 (Hidayat et al., 1996; Hidayat and Ishii, 1998, 1999; Berger and Zoro, 2004). The first LLS to operate in Malaysia was installed in 1996 and was composed of 8 MDF sensors (Yahaya et al., 1996). However, data of this LLS were not included in this book because it is not believed to be reliable (Abidin and Ibrahim, 2003). Other LLS was installed in Malaysia in 2003, replacing five old MDF sensors by 5 IMPACT sensors (Abdullah and Yahaya, 2008).

One more LLS operated in the tropical region for a short period: the three-sensor ALDF LLS installed in Papua New Guinea (Orville et al., 1997). Data from this network were not included in the comparative analysis that will be presented related to the maximum ground flash density because the maximum flash density observed in Papua New Guinea occurs outside the region limited by the sensors. However, mean values of the other parameters were considered.

Two other networks cover very small areas on the edge of the tropical region and the results obtained in the tropical region were considered: South China (Chen et al., 2002, 2004) and South Africa (Ndlovu and Evert, 2004; Evert and Schulze, 2005; Bhikha et al., 2006; Gill, 2008).

Table 4.1 summarizes the main characteristics of the LLS in the tropical region, indicating the countries where there are/were observations. It includes the period of the data, number and type of sensors, detection technique, approximate latitude range of the observations in the tropical region and detection efficiency. It can be observed from this table that, even though only a small fraction of the tropical region is covered by LLS, almost all latitudes were partially covered. Data from Brazil and Java Island were divided in different data sets corresponding to different networks, regions and/or periods, as described before.

Table 4.1. Characteristics of LLS in the tropical region

Country	Period of Data	Number of Sensors	Type of Sensors	Detection Technique	Latitude Range	Average Flash Detection Efficiency (%)
Australia - North	1998-2003	16	LPATS	TOA	11°-23.5° S	70[1]
Brazil Southeast-1	1988-1996[2]	6	LPATS	TOA	16°-22° S	70[1]
Brazil Southeast-2	1997-2005	24	LPATS/IMPACT	TOA/MDF	13°-23.5° S	90[3]
Brazil -North	1999-2005	4	ALDF	TOA/MDF	8°-14° S	90[1]
Brazil - BrasilDat	2007-2008	39	LPATS/IMPACT	TOA/MDF	0°-23.5° S	70-90[4]
Colombia	1995-2003	6	LPATS	TOA	3°-11° N	70[1]
Java Island-1	1994-1995	4	ALDF	TOA/MDF	6°- 9° S	70[1]
Java Island-2	1996-2001	8	LPATS	TOA	6°-9° S	60[5]
Malaysia	2004-2007	8	ALDF/IMPACT	TOA/MDF	1°-7° N	90[1]
Papua New Guinea	1992-1994	3	ALDF	TOA/MDF	5° S – 2° N	70[1]
Venezuela	2000-2003	12	IMPACT	TOA/MDF	0°-12° N	90[1]

[1] Value was given by the manufacturer.

[2] The number of sensors changed from 4 to 6 in 1995 (Pinto et al., 2003b).

[3] Value was estimated by high-speed camera observations (Ballarotti et al., 2006).

[4] Lower value corresponds to that given by the manufacturer for the North region. The upper value corresponds to the Brazil Southeast-2 LLS value, as well as that given by the manufacturer for the South region.

[5] Value was estimated from the lowest peak current observed.

Table 4.1 indicates that only for the Brazil Southeast-2 LLS the detection efficiency was validated by independent observations. Solorzano (2003) estimated from a small sample of triggered flashes obtained in 2001 and 2002 a stroke detection efficiency of 50% for the Brazil Southeast-2 LLS. Later on, Ballarotti et al. (2006) estimated the stroke and flash detection efficiency for the Brazil Southeast-2 LLS based on the observation of 233 negative ground flashes in 2003 and 2004 using a high-speed camera. They found a stroke detection efficiency of 55% and a flash detection efficiency of 88%. These values are in reasonable agreement with the values obtained in Florida for the National Lightning Detection Network -

NLDN (Jerauld et al., 2005). For BrasilDat LLS the detection efficiency was validated only for the Southeast. For the South and North regions, the only detection efficiency values were those provided by the manufacturer. In the case of Java Island-2, the detection efficiency in Table 1 was estimated from the lowest peak current observed. For the other LLS, the values of the detection efficiency in Table 1 were provided by the manufacturers and are based on a model that assumes a perfect operation of the system. In this sense, they should be considered as achievable values and might differ from real values.

4.3. DETECTION EFFICIENCY MODELS FOR LLS

Like any other detection system, an LLS has its limitations (Cummins and Bardo, 2004; Naccarato and Pinto, 2008; Diendorfer et al., 2008b). Perhaps the most significant one is its detection efficiency (DE), which is the ratio of the number of detected events divided by the real number of events. Due to the high variability of the ground lightning physical features, an LLS will never be able to detect all events, thus its DE will never reach 100%. Such errors can be more or less significant depending on the frequency of sensor faults, communication problems or the sensor network geometry, which may be unfavorable (Schulz, 1997; Cummins et al. 1998a, b; Naccarato et al., 2004a; Naccarato et al, 2006b), leading to distortions in the data. Furthermore, the type of sensors plays a very important role. The main purpose for evaluating the DE of an LLS is to separate the geographical variations of the ground lightning parameters from the variations due to the LLS performance (Cummins and Bardo, 2004).

Several approaches for developing a relative detection efficiency model (RDEM) have been reported in the literature: Schulz (1997), Murphy et al. (2002), Rompala et al. (2003) and Naccarato et al. (2004b). All of them use a set of downward ground lightning data reported by the network to compute its relative detection efficiency (RDE). Since the lightning dataset used as reference is usually collected over areas with higher values of DE (known from other independent observations), the resulting RDE can be roughly approximated to the absolute detection efficiency (ADE).

Cummins et al. (1992, 1993) and Schulz (1997) were the first to publish a comprehensive description and a clear methodology for an absolute detection efficiency model (ADEM) based on data from the NLDN in the United States and from the LLS in Austria, respectively.

Schulz (1997) presents a model to estimate the network first-stroke detection efficiency (FSDE), allowing the correction of the peak current distributions based on the data provided by the same network. The stroke detection efficiency (SDE) is defined in terms of individual strokes. The relation between SDE and the flash detection efficiency (FDE) depends strongly on the distribution of the number of strokes per flash, and the FDE is always higher than the SDE (Rubinstein, 1995). According to Schulz (1997), to estimate the FSDE of a given LLS for a specified peak current range it is required to assume a sensor DE function. Then, using these individual sensor DE functions, the entire network DE is computed based on the combined probability of each sensor to detect a first-stroke, considering its peak current intensity and the distance from the sensor. Of course, the propagation effects are intrinsically taken into account, since the particular DE of each sensor is computed based on first-stroke data detected by the network for different distances.

The first results regarding an absolute flash detection efficiency model (AFDEM) for the NLDN were published by Cummins et al. (1992, 1993, 1995). The model computed the network FDE on a 50 km x 50 km grid over the coverage area. At each grid point, the model generated specific values of peak current and computed the signal strength that should arrive at each sensor in the network using a signal propagation model. The model then used a look-up table to relate the computed signal strength at each sensor to its FDE, thus providing the probability for the flash to be detected by the sensor. The look-up table contained the response of each type of lightning sensor as a function of the incident signal amplitude, and the values ranged from zero probability at threshold to a maximum probability (less than one) at 2-3 times threshold. Finally, the final network FDE was computed for all combinations of sensors that reported a discharge, assuming that the sensor probabilities are all independent.

Later on, Murphy et al. (2002) used a very simple methodology to estimate the improvement in the NLDN SDE due to the 2002 NLDN upgrade. Two sets of stroke data were analyzed: one before and one after the upgrade for the same area and time period. The cumulative peak current distribution (PCD) was then computed for the two datasets and the early PCD (before the upgrade) was then fitted into the new PCD (after the upgrade), which was considered the reference (100% SDE). The ratio between the reference PCD and the fitted PCD represented the RSDE improvement of the network.

Rompala et al. (2003) developed a method to estimate the SDE contours for the Brazil North LLS. They first selected an area in the middle of the network (composed only by 4 IMPACT sensors), which was assumed to have the best performance. This region was called central quad. The PCD for the downward ground stroke data in the QUAD was computed and adjusted to a theoretical probability distribution function (PDF), which was considered to be representative of the stroke distribution at any location over the region. After that, the network coverage area was divided into blocks (cells) of a specific size and the PCD was computed for every stroke detected in each cell. The reference PCD was then applied to each cell to assess which proportion of the lightning set would be detected. Finally, the cell SDE was taken as the ratio of the computed values divided by the total number of events.

Naccarato et al. (2004b) combined the Murphy et al. (2002) and Rompala et al. (2003) methodologies described above, to develop a quite simple technique to assess the relative flash detection efficiency (RFDE) of LLS. It showed a relatively good agreement with the expected network behavior. Of course, like any other method that requires lightning data, this approach was highly dependent on the number of detected events to provide good results. Thus, the higher the number of events, the longer the calculation time. In order to ensure good statistics, the number of flashes for each cell was calculated as the sum of the events of the 8 surrounding cells and the cell itself. Finally, the PCD for each cell was computed individually and adjusted to the reference PCD (Murphy et al. 2002). Figure 4.4 presents the resulting RFDE map for the methodology described above for the Brazil Southeast-2. It can be inferred that this methodology is able to identify the areas of maximum DE and also to describe the continuous decrease of the DE far away from the sensors. For example, for this approach, regions outside the network presented DE values lower than 55%.

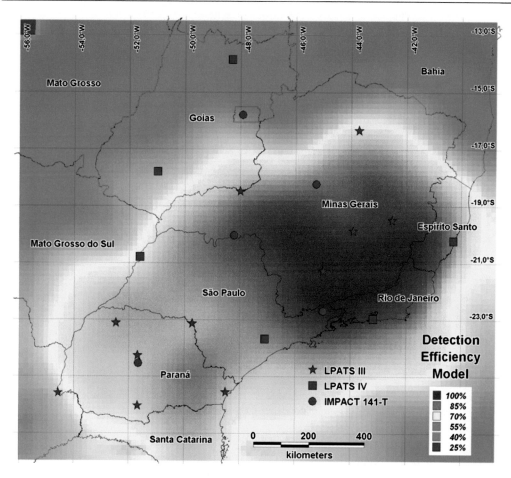

Figure 4.4. Relative stroke detection efficiency of the Brazil Southeast-2 LLS for a 22-sensor hybrid network. The grid resolution is 50 km x 50 km (adapted from Naccarato et al., 2004b).

In order to overcome the limitations of the methodology described in Naccarato et al. (2004b), Naccarato et al. (2006a) developed a RFDEM that also requires the stroke data detected by the network. However, they now used them only to compute the RSDE distribution function for each sensor, which depends on the peak current and the distance from the stroke due to the propagation effects. Thereby, using these individual sensor RSDE curves, the network RSDE is computed based on the combined probability of each sensor to detect (or not) the stroke. This new approach employs a physical model (Schulz, 1997) and reduces significantly the calculations. This allows an easy evaluation of the network DE due to changes in its sensor geometry (by enabling or disabling specific sensors or including new virtual sensors). Moreover, this new approach considers that a valid stroke solution can be achieved by both the minimum required number of angles (IMPACT sensors) and/or times (IMPACT and/or LPATS sensors) or by only two IMPACT reports (i.e., two bearing and two timing information). Figures 4.5 and 4.6 show the RFDE for the Brazil Southeast-2 network for a 10 km x 10 km spatial resolution computed for different peak current ranges.

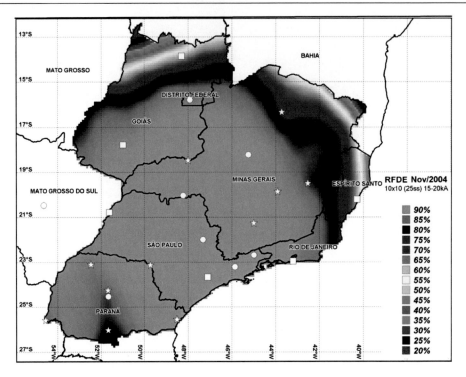

Figure 4.5. Relative flash detection efficiency of the Brazil Southeast-2 network computed for 15-20 kA peak current range in Nov. 2004, when all 25 sensors were operating. The grid resolution is 10 km x 10 km (adapted from Naccarato et al., 2006a).

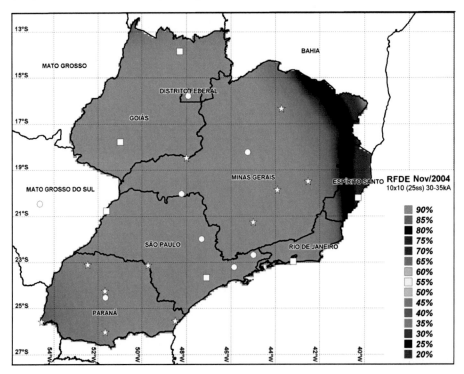

Figure 4.6. Same as Figure 4.5, but for 30-35 kA peak current range (adapted from Naccarato et al., 2006a).

As discussed previously, the main goal of correcting the ground flash density maps is to minimize the relative geographical variation in the ground lightning data associated to variations in the network DE. Thus, after the correction, the relative differences in the ground flash density values are expected to be mostly due to physical factors rather than LLS performance. Figure 4.7 shows the original ground flash density map (without correction) for southeastern Brazil considering a 6-year ground lightning flash dataset (1999-2004). Figure 4.8 shows the corrected flash density values for the Brazil Southeast-2 for the same period in Figure 4.7.

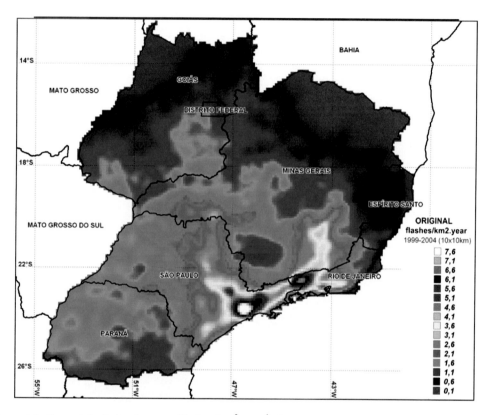

Figure 4.7. Ground flash density map (flashes.km^{-2}.year^{-1}) for a 6-year dataset (1999 2004) obtained by the Brazil Southeast-2 network without correction of RFDE. The grid resolution is 10 km x10 km (adapted from Naccarato et al., 2008).

Comparing Figures 4.7 and 4.8, Naccarato et al. (2008) observed that the correction of the flash density values works effectively only for regions with network RFDE values between 60% and 80%. Below 60% RFDE, the correction is not effective due to the reduced number of ground strokes detected by the network, leading to ground flash densities smaller than expected. They arrived at this conclusion comparing the ground lightning data with total lightning data provided by the LIS sensor for the same region and time period. On the other hand, above 80% DE, the model correction does not increase appreciably the number of CG flashes, which prevents it from recovering the actual values of flash densities, again comparing the ground lightning data to LIS data.

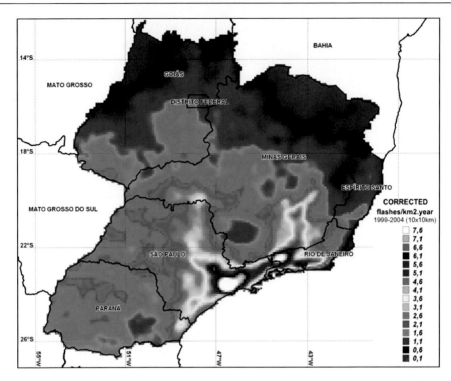

Figure 4.8. Same as Figure 4.7, for the corrected ground flash density values in flashes.km^{-2}.year^{-1} (adapted from Naccarato et al., 2008).

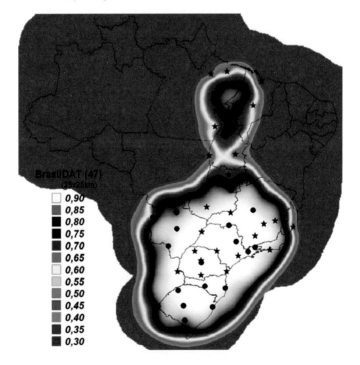

Figure 4.9. BrasilDat network RFDE computed by the Naccarato et al. (2008) model using the original sensor RSDE curves for all solutions. The spatial resolution is 25 km x 25 km (adapted from Naccarato et al., 2008).

To make the results more realistic closer to the network boundaries, where most of the sensors are essential, the solution is computed with a minimum number of reporting sensors, Naccarato and Pinto (2008) developed a new model that neglects such solutions in the computation of the RFDEM. The effect of their model can be seen comparing Figures 4.9 and 4.10. Figure 4.9 presents the BrasilDat network RFDE computed by the RFDEM using the sensor RSDE curves that include all solutions, while Figure 4.10 shows the BrasilDat network RFDE considering the curves neglecting the unique solutions. The magenta represents RFDE values below 30%. The white color represents RFDE values up to 88%. From both figures, it can be observed that the network has a very high performance in mid-southern Brazil, decreasing significantly towards the north due to the small network composed of only LPATS sensors. In general, almost 40% of the country is in fact covered by a network with about 80-90% RFDE. One can observe that the "border effect" decreases clearly since the RFDE values calculated by the new methodology are lower throughout the network boundaries. On the other hand, as expected, the estimated RFDE within the network was not affected by the removal of the unique solutions in the sensor RSDE curves calculation.

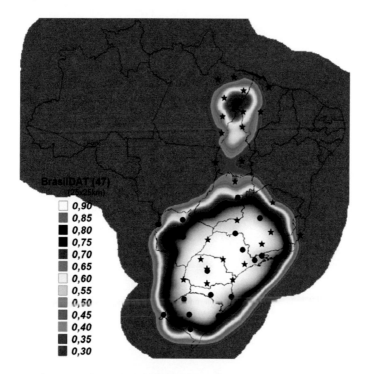

Figure 4.10. Same as Figure 4.9, but using the Naccarato and Pinto (2008) RSDE model neglecting the "essential" solutions (adapted from Naccarato et al., 2008).

Figure 4.11 shows the ground flash density values based on a 2-year flash data (from Jun/2005 to May/2007) provided by BrasilDat network without the correction regarding the FDE variations and Figure 4.12 presents the corresponding corrected flash density values. Based on the comparison of these figures, Naccarato and Pinto (2008) suggest that their BrasilDat RFDEM model presents a reasonable sensitivity that allows testing any LLS for unfavorable sensor geometries and/or looking for regions with bad or insufficient sensor coverage.

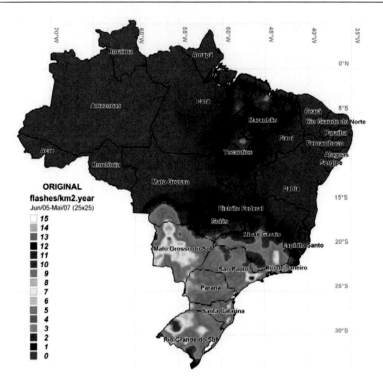

Figure 4.11. Ground flash density values (flashes.km^{-2}.year^{-1}) for a 2-year BrasilDat dataset (Jun/2005 to May/2007) without any correction in the variations of the network RFDE. The grid resolution is 25 km x 25 km (adapted from Naccarato and Pinto, 2008).

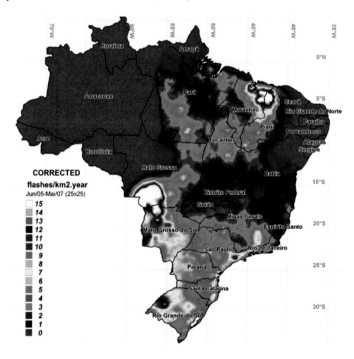

Figure 4.12. Same as Figure 4.11, but for corrected flash density values in flashes.km^{-2}.year^{-1} (adapted from Naccarato and Pinto, 2008).

4.4. OBSERVATIONS

4.4.1. Flash Density

This section concerns only lightning flash density observations over the continents, since LLS observations over the oceans are out of their best performance region. As recently documented (Murphy and Holle, 2005), observations very distant from the sensors, such as those over oceans, should be seen with caution when not adequately corrected for the detection efficiency of the LLS.

Table 4.2. Maximum ground flash density and respective average percentage of positive ground flashes observed by LLS in the tropical region.

Country	Maximum Ground Flash Density[1] (flashes.km^{-2}.year^{-1}) (Reference)	Spatial Resolution (km x km)	Period of Observation	Percentage of Positive Ground Flashes (%) (Reference)	Physical-related Process
Australia - North	12 (Kilinc and Beringer, 2007)	1x1	1998-2003	75 (Kilinc and Beringer, 2007)	Moonsoon/Mountain Interactions
Brazil Southeast-1	9 (Pinto et al., 2004c)	9x9	1988-1996	35 (Pinto et al., 1999a)	Cold Front/Mountain Interactions
Brazil Southeast-2	24-25 (Naccarato, 2005)	1x1	2001	12 (Naccarato, 2005)	Cold Front /Urban Area Interactions
	17-18 (Naccarato, 2005)	1x1	1999-2004		
	12-13 (Naccarato, 2005)	4x4	1999-2004		
	9-10 (Naccarato, 2005)	10x10	1999-2004		
Brazil - North	12 (Fernandes, 2005)	10x10	2002-2003	20 (Blakeslee et al., 2003)	ITCZ[2] /Amazon forest Interactions
	5 (Fernandes, 2005)	55x55	2002-2003		
Brazil - BrasilDat	10 (Naccarato et al., 2008)	25x25	2007-2008	12 (Naccarato, 2005)	Cold Front /Urban Area Interactions
Colombia	35 (Younes et al., 2004)	3x3	1997-2001	70	ITCZ/ Mountain Interactions
	47 (Younes et al., 2004)	3x3	1997	(Younes et al., 2003)	
	16-17 (Younes et al., 2004)	30x30	1997-2001		
Java Island-1	16 (Hidayat and Ishii, 1998)	12x12	1995	-	ITCZ /Sea-land Interactions
Java Island-2	39 (Berger and Zoro, 2004)	1x1	1999	20 (Berger and Zoro, 2004)	ITCZ /Sea land Interactions
	25 (Berger and Zoro, 2004)	8x8	1999		
Malaysia	12 (Abdullah and Yahaya, 2008)	4x4[3]	2004-2007	20 (Abdullah and Yahaya, 2008)	ITCZ /Sea-land Interactions
Papua New Guinea	-	-	1992-1994	6 (Orville et al., 1997)	ITCZ /Sea-land Interactions
Venezuela	34-39[4] (Tarazona et al., 2006)	3x3	2000-2003	23 (Tarazona et al., 2006)	ITCZ/ Mountain Interactions

[1] Values were not corrected for detection efficiency of the LLS.

[2] Intertropical Convergence Zone.

[3] Data not available. It was estimated between 3 and 5 km.

[4] Values were calculated from the stroke density, assuming the negative multiplicity between 2 and 2.5 and the positive multiplicity equals one, typical values for IMPACT LLS.

Figures 4.7 and 4.11 shown previously are examples of maps of the annual average ground flash density for the Brazil Southeast-2 and Brazil-BrasilDat networks. Similar maps for the other LLS in the tropical region indicated in Table 4.1 can be found in: Younes et al. (2004) for Colombia, showing a maximum ground lightning flash density in the region around 8° N of latitude and 75.5° W of longitude; Tarazona et al. (2006) for Venezuela, showing a maximum ground lightning flash density in the region around 10° N and 72° W; Hidayat and Ishii (1998) for Java Island-1, showing a maximum ground lightning flash density in the region around 7° S and 107.5° E; Berger and Zoro (2004) for Java Island-2, showing (in a tabular form) a maximum ground lightning flash density in the region around 6.5° S and 108° E; Pinto et al. (2003b) for Brazil Southeast-1, showing a maximum ground lightning flash density in the region around 20° S and 44.5° W; and Fernandes (2005) for Brazil North, showing a maximum ground lightning flash density in the region around 11° S and 62.5° W.

Table 4.2 shows a summary of the maximum ground flash densities observed by the different LLS described in Table 4.1. Values are not corrected for the detection efficiency of LLS and, when available, are presented for different spatial resolutions. The periods of the observations are also shown. Ground flash densities were obtained by grouping strokes in flashes using different criteria. For LPATS LLS, in general, strokes were grouped assuming that all strokes belong to the same flash if they are located less than 10 km from the first stroke and occur less than two seconds from the first stroke, since at the same time consecutive strokes are separated by less than 500 ms. For hybrid (LPATS and IMPACT sensors) LLS or IMPACT LLS the criteria normally used is the one suggested by Cummins et al. (1998a). Extreme values for specific years (1997 in Colombia, 1999 in Java Island and 2001 in Southeastern Brazil) are also included in Table 4.2. The values in Table 4.2 vary from 5 to 47 flashes km^{-2} $year^{-1}$. In general, the higher the spatial resolution, the higher the maximum lightning density. The maximum value observed for all LLS in Table 4.2 was in 1997 in Colombia, for a spatial resolution of 3 km x 3 km.

Table 4.2 also indicates the percentage of positive CG lightning flashes and the physical processes related to the maximum CG lightning flash densities observed by the different LLS. They include different meteorological systems and their interaction with different geographical features. In North Australia, the maximum ground flash density is related to the monsoon interaction with local mountains. In Colombia, the maximum ground flash density occurs in the North part of the country and is related to the seasonal variation of the trade winds associated with the oscillation of the Intertropical Convergence Zone (ITCZ) and its interaction with local mountains. In Southeast Brazil the maximum CG lightning flash density is related to the interaction of cold fronts and local mountains, in the case of Brazil Southeast-1, while in Brazil Southeast-2 (and Brazil-BrasilDat) it is related to the interaction of cold fronts with the urban area of São Paulo in the presence of sea breeze. In the north region of Brazil, although the presence of the Amazon forest has a considerable effect (Williams et al., 2002), the maximum ground flash density is also related to the seasonal variation of the trade winds associated with the oscillation of the ITCZ. In the Java Island, the maximum ground flash density is related to the seasonal variation of the trade winds associated with the oscillation of the ITCZ and the land/ocean interaction. In Malaysia, the situation is the same as in the Java Island. Finally, in Venezuela, the maximum ground lightning flash density occurs in the northwest part of the country, in a region close to the region of maximum ground lightning flash density in Colombia, and is related to the same processes that occur in Colombia.

In order to make a more reliable comparison among the values of maximum ground flash densities in different countries presented in Table 4.2, Pinto et al. (2007) claimed that three aspects should be considered: different spatial resolutions of the observations, different detection efficiency of each LLS and different contamination of each LLS by cloud flashes. With respect to the spatial resolution, as observed before, the higher the spatial resolution, the higher the maximum lightning density. In consequence, in order to compare values at different spatial resolutions, it is necessary to know in details the spatial distribution of the flash density in the different countries. Since this information is not available for most countries, this effect can be estimated taking the available information of maximum ground flash densities for different spatial resolutions in Colombia, Java Island-2, Brazil North and Brazil Southeast-2 and, then, extrapolated to all countries. Figure 4.13 shows the variation of the maximum ground flash density versus spatial resolution taking the values in Brazil Southeast-2 as reference. The values in Colombia, Java Island-2 and Brazil North were normalized to the values in Brazil Southeast-2 for the lower spatial resolutions available in these countries. Data from the different countries in Figure 4.3 can be fitted into a logarithm curve with a square correlation coefficient (R^2) equal to 0.91.

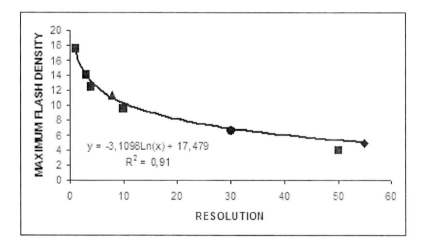

Figure 4.13. Variation of the maximum ground flash density in flashes.km^{-2}.year^{-1} as a function of the resolution in km x km. The results are based on data from the LLS in Brazil Southeast-2 (black square), Brazil North (black rhombus), Colombia (black dot) and the Java Island-2 (black triangle).

The detection efficiency of LLS, as discussed before in this chapter, depends on several aspects including distance and type of sensors, network configuration, site characteristics, and others. For the LLS in Table 4.1, the detection efficiency varies from 60% in Java Island-2 to 90%. For the LLS the values of detection efficiency in the region of maximum ground flash density are available (Java Island-1, Brazil Southeast-1, Brazil Southeast-2, Brazil-BrasilDat and Brazil-North). They were used to correct the data. Otherwise, average values inside the LLS were used.

Following Pinto et al. (2007), the last aspect that should be considered when comparing data from different LLS is the possible influence of the cloud flash contamination on the maximum ground flash densities. The contamination is a result of misclassification of cloud flashes mainly as low peak current positive flashes. The resultant effect is an increase in the maximum ground lightning flash density. In order to investigate this aspect, the values of the percentage of positive ground flashes shown in Table 4.2 must be considered. For all LLS this percentage is around 10%-20%, except for the LPATS LLS in Brazil Southeast-1, Colombia and Australia-North, where they are 35%, 70% and 76%, respectively. In the case of Brazil Southeast-1, later observations also using IMPACT sensors (Brazil Southeast-2) in the same region showed that this large percentage is a result of cloud contamination. This agrees with observations outside the tropical region which suggest that in this type of LLS positive ground flashes are more contaminated by cloud flashes (Cummins et al., 1998a, b; Théry, 2001; Rakov and Uman, 2003). In the case of Colombia and Australia-North, there are no other reliable observations to clarify this issue. However, considering the results obtained in Brazil Southeast-1 and that the large percentages of positive ground lightning observed in these countries are not supported by indirect indications obtained from sprite and mesoscale convective system distributions (Toracinta and Zipser, 2001; Füllekrug and Price, 2002; Sato and Fukunishi, 2003; Cecil et al., 2006), it is assumed that they are probably the result of cloud flash contamination.

Table 4.3. Absolute maximum ground flash densities observed in the tropical region.

Country	Period of Observation	Absolute Maximum Ground Flash Density[1] (flashes.km^{-2}.year^{-1})
Australia-North	1998-2003	9
Brazil Southeast-1	1988-1996	19
Brazil Southeast-2	1999-2004	21
	2001	29
Brazil-North	2002-2003	34
Brazil-BrasilDat	2007-2008	29
Colombia	1997-2001	34
	1997	46
Java Island-1	1995	33
Java Island-2	1999	65
Malaysia	2004-2007	20
Venezuela	2000-2003	56

[1] Values refer to a spatial resolution of 1 km x 1 km.

The aspects mentioned above were then applied to the values of maximum ground lightning flash densities shown in Table 4.2. First, the values of maximum ground flash densities in Table 4.2 were normalized to the spatial resolution of 1 km x 1 km, following the equation shown in Figure 4.13. After that, the values were corrected for variations in the detection efficiency, following the values described in Table 4.1. Finally, the values in Brazil Southeast-1, Colombia and Australia-North were adjusted considering that their percentages of positive flashes are equal to those of Brazil Southeast-2 and Venezuela, respectively. This approach is based on the proximity of the regions and on the fact that the LLS in Brazil Southeast-2 and Venezuela are less subject to cloud flash contamination due to the fact that they include IMPACT sensors. In the case of Venezuela the value may still be high considering some local observations (Falcon et al., 2001, 2007). In the case of Australia-North the value of 20% was adopted. The resultant values were named absolute maximum ground flash densities and are shown in Table 4.3. The main result of this table is that the absolute maximum ground flash density in the tropical region ranges from 21 to 65 flashes.km^{-2}.year^{-1}. The largest value occurs in Java Island-2 and not in Colombia as was the case for the maximum ground flash densities observed.

Regarding the values in Table 4.3, two other statistical uncertainties should be considered. The first is related to the effect of the location accuracy on the maximum flash density. This topic has been discussed by Campos and Pinto (2007), who found that for LLS with a typical mean location accuracy of 1 km the effect on the absolute maximum flash density value is of about 10%. The second is related to the variance of the number of flashes observed in each grid cell (Diendorfer, 2008). This variance increases as the number of total events in each grid cell decreases. For the values in Table 4.3, it can be as high as 30%, for the observations related to just one year.

4.4.2. Peak Current

The estimation of stroke peak current by LLS is based on a linear relationship between peak electromagnetic field and peak current. The relationship has been obtained experimentally comparing peak current estimated by LLS with triggered lightning peak current measurements (Orville, 1991; Idone et al., 1993; Jerauld et al., 2005; Diendorfer et al., 2008a, b; Nag et al., 2008) and is believed to be valid for subsequent negative strokes of downward negative ground flashes. For first-stroke of negative downward flashes and positive downward flashes no observations exist, so the relationship should be considered with caution. The relationship is supported by a simple transmission line model (Uman et al., 1975) in which a constant return stroke velocity is assumed, the ground is perfectly conducting and the top of the channel is at the infinite.

Due to the high variability of some key parameters such as the return stroke speed (Rakov, 2007) and propagation effects, it is not possible to determine the lightning current accurately from the remotely measured electric or magnetic field for any given event. Nevertheless it has been shown by Rachidi et al. (2004) that the transmission line model relationship is also valid when considering mean values of peak field, peak current and return stroke velocity. This is, to some extent, a theoretical justification to use lightning location systems to infer lightning current statistical distributions from measured fields alone.

However, before the peak fields are converted to peak current values, propagation effects must be taken into account to produce a range-normalized value of the field signal for each sensor (Cummins et al., 1998a). Different attenuation models have been used to take into account the propagation effects (Orville, 1991; Idone et al., 1993; Herodotou et al., 1993; Cramer et al., 2004; Diendorfer, 2007). The final peak current estimates are a combination of individual values for each sensor, considering some restrictions that may involve the type of sensors in hybrid LLS and/or the distance of the sensor to the lightning location.

The estimation of the mean peak current by LLS, however, is considerably affected by the low stroke detection efficiency of the LLS for low peak current subsequent strokes (Schulz and Diendorfer, 1998; Cummins and Bardo, 2004; Pinto, 2008b), by the cloud flash contamination mainly in LPATS LLS and by positive ground flashes (Cummins and Bardo, 2004). In addition, LPATS LLS are also very sensitive to sensor calibration. For this reason, in the majority of the LLS, mean peak current values are considerably overestimated, making it very difficult to know if regional differences in this parameter are real or if they are mostly due to variations in the LLS performance (Pinto et al., 2004b, 2007).

Table 4.4 shows average peak current values for negative and positive ground flashes obtained by LLS in the tropical region, considering that it is represented by the peak current of the first-stroke. In some cases, positive values are filtered above a fixed threshold to minimize cloud flash contamination.

**Table 4.4. Average peak currents for first-stroke ground flashes observed by LLS i
n the tropical region.**

Country	Negative Peak Current (kA) (Reference)	Positive Peak Current (kA) (Reference)
Australia	39[1] (Peterson and Rutledge, 1992)	39[1] (Peterson and Rutledge, 1992)
Brazil Southeast-1	40 (Pinto et al., 2003b)	39[2,3] (Pinto et al., 1999a)
Brazil Southeast-2	23 (Pinto et al., 2008)	29[4] (Pinto et al., 2008)
Brazil North	33 (Blakeslee et al., 2003) 27 (Fernandes, 2005)	21 (Blakeslee et al., 2003)
Colombia	-	-
Java Island-1	37 (Hydayat et al., 1996)	41 (Hydayat et al., 1996)
Java Island-2	40-55 (Berger and Zoro, 2004)	-
Malaysia	37 (Abdullah and Yahaya, 2008)	-
Papua New Guinea	25[2] (Orville et al., 1997)	33[2] (Orville et al., 1997)
Venezuela	13[2] (Tarazona et al., 2006)	-

[1] From a 4-MDF LLS during the 1989-1990 summer season.
[2] Median values.
[3] Values refer to peak currents above 15 kA.
[4] Values refer to peak currents above 10 kA.

Peak current values in Table 4.4 vary from 13 kA to almost 55 kA. The variation seems to be more related to the performance of the different LLS than geographical changes, even though Biagi et al. (2007) have suggested that peak current observed by the NLDN (and validated by video in terms of ground flashes) may change from storm to storm.

4.4.3. Multiplicity

Various methods can be used to group strokes into flashes and determine the flash multiplicity from LLS data, affecting the resulting value. Older LLS, based only on the MDF sensors, employed an angle-based algorithm where each MDF sensor counted all strokes that occurred within an interval of about ±2.5 degrees of the first stroke for a period of one second after the first stroke. The assigned flash multiplicity was simply the largest number of strokes detected by the MDF sensors, assuming an upper limit to the multiplicity. Other grouping algorithms group strokes into flashes using a spatial and temporal clustering algorithm illustrated in Figure 4.14. Strokes are added to any active flash for a specified time period (usually 1 second) after a first stroke, as long as the additional strokes are within a specified clustering radius (usually 10 km) of the first stroke and the time interval from the previous stroke is less than a maximum interstroke interval (usually 500 ms). Additionally, in modern central processors, if a stroke is located farther than the clustering radius from the first stroke but is not clearly separated from it because their location confidence regions overlap, then the stroke is included in the flash. Depending on the system configuration, strokes may be counted in the multiplicity even if they have a polarity that is opposite from that of the first stroke.

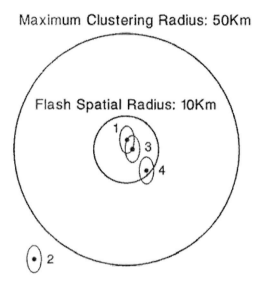

Figure 4.14. Spatial clustering method for grouping strokes into flashes.

Table 4.5. Multiplicity of negative ground flashes observed in the tropical region.

Country	Negative Multiplicity or percentage of single strokes (Reference)
Brazil Southeast-1	56%[1] (Pinto et al., 2003b)
Brazil Southeast-2	1.9 (Pinto et al., 2008)
Brazil North	65%[1] (Blakeslee et al., 2003)
Brazil BrasilDat[2]	1.6 (Bourscheidt et al., 2008)
Colombia	1.5 (Torres et al., 2008)
Malaysia	3 (Abdullah and Yahaya, 2008)
Papua New Guinea	2.3 (Orville et al., 1997)

[1] Only percentage of single strokes was calculated.

[2] Values correspond to the South region. In the Southeast region, they are the same as Brazil Southeast-2.

Almost the same conditions mentioned with respect to the estimates of peak current by LLS apply to observations of multiplicity, since it is considerably affected by the low stroke detection efficiency of the LLS for low peak current subsequent strokes (Schulz and Diendorfer, 1998; Cummins et al., 1998a; Rakov and Uman, 2003). Contamination by the cloud flashes may also affect considerably the multiplicity in general adding false single stroke flashes. For these reasons, in the most LLS, mean multiplicity values are considerably underestimated, making it very difficult to know if regional differences in this parameter are real or due to LLS performances (Pinto et al., 2004b, 2007). Table 4.5 shows the observed multiplicity of negative ground flashes in different countries in the tropical region. Only negative multiplicity is shown since positive ground flashes are mostly single flashes.

All multiplicity (percentage of single flashes) values in Table 4.5 are lower (higher) than the typical values found in accurate-stroke studies (Rakov and Uman, 2003; Ballarotti et al., 2008), which suggest that LLS estimates of multiplicity are not accurate and underestimate the real values of multiplicity.

4.4.4. Comparison with Total Lightning Observations by Satellites

In this section, maximum ground lightning flash densities observed by the different LLS in the tropical region are compared to total lightning densities obtained by OTD and LIS satellite observations. The comparison is based on the values of the cloud flash to ground flash ratio, or cloud to ground ratio for short. This ratio, in turn, is associated with the thunderstorm characteristics (Williams et al., 1999; Souza et al., 2008).

Table 4.6 shows an estimate of the cloud to ground ratio for the same regions of the maximum ground flash densities shown in Table 4.3. The values in Table 4.3 were converted to the spatial resolution of the total lightning data (approximately 55 km x 55 km) using the curve in Figure 4.4. Only the LLS that operated for at least two years after 1995 (when OTD was launched) were considered in the comparison, due to the limitations in the satellite sample rate. The values of cloud to ground ratio obtained for the regions of maximum ground flash densities observed by the different LLS in the tropical region in Table 4.6 show large variations, ranging from 3.9 to 12.6.

Table 4.6. Total lightning densities observed in the tropical region by OTD and LIS and the inferred cloud to ground ratios.

Country	Total Lightning Flash Density[1] (flashes km^{-2} year^{-1})	Cloud to Ground Ratio[2]
Australia-North	20	6.6
Brazil Southeast-1	20	4.0
Brazil Southeast-2	26	4.9
Brazil-North	25	3.9
Brazil-BrasilDat	26	4.2
Colombia	99	11.0
Malaysia	35	6.9
Venezuela	181	12.6

[1] Values were obtained by a data set kindly provided by Dr. Steve Goodman from Marshall Space Flight Center. They were corrected by variations in the view time and detection efficiency of the satellite sensors.
[2] Values were calculated for a spatial resolution of approximately 55 km x 55 km.

4.4.5. Comparison with LLS Observations in the Temperate Region

In this section, the absolute maximum ground flash density, mean peak current, multiplicity and percentage of positive ground flashes, as well as monthly lightning distribution of these parameters, observed by LLS in the tropical region are compared with similar data obtained in the temperate region. Due to the large number of LLS in the temperate region (more than 30), the comparison of absolute maximum ground flash densities was limited to consider only the highest values observed by LLS in each continent of the temperate region, and the comparisons involving peak current, multiplicity and percentage of positive ground flashes were limited to consider only three LLS located in Brazil, the United States and Austria, which are believed to be the more reliable networks operating for more than a decade in the tropical and temperate regions, respectively.

Table 4.7 shows the highest values observed in each continent: in North America, the highest value was observed in Florida, United States (Orville et al., 2002; Murphy and Holle, 2005); in Europe, the highest value was observed in the Northwest region of Italy (Schulz et al., 2005); in Africa, the highest value was observed in the North region of South Africa (Ndlovu and Evert, 2004; Evert and Schulze, 2005; Bhikha et al., 2006; Gill, 2008); and, finally, in Asia the highest value was observed in the South region of China (Chen et al., 2002, 2004).

Table 4.7. Maximum ground flash density observed in different continents in the temperate region.

Continent (region)	Maximum Ground Flash Density[1] (flashes/km².year) (Reference)	Spatial Resolution (km x km)	Period of Observation
North America (Florida, United States)	16 (Murphy and Holle, 2005)	1x1	1996-2000
	9-12 (Orville et al., 2002)	20x20	1989-1998
Europe (Northwest region of Italy)	15 (Schulz et al., 2005)	1x1	1992-2001
	6-7 (Schulz et al., 2005)	20x20	1992-2001
Asia (South region of China)	7-8 (Chen et al., 2002)	30x30	1997-2001
Africa (North region of South Africa)	19[2,3]	1x1	2001-2003

[1] Values were not corrected for the detection efficiency of the LLS.

[2] Data were obtained from SKA Site Selection South Africa, Report to the ISSC, Eskom, Dec. 2003. The number of flashes were obtained considering that strokes are part of a flash if the distance between successive strokes is less than 2.5 km and time between successive strokes is less than 100 ms.

3 Data obtained in the 2006-2007 period from Gill (2008) based on the new LS7000 LLS were not considered, since the observations were not corrected for cloud contamination which has been shown to be significant for this type of sensors (Naccarato and Pinto, 2008).

Table 4.8 shows the absolute maximum ground flash density calculated from the values in Table 4.7. The following detection efficiency values, taken from the articles cited in the previous paragraph, were used to correct the observed densities: 90% for the regions in the United States and Italy, 85% for South China and 80% for the North of South Africa. By comparing Tables 4.3 and 4.7 it can be observed that the absolute maximum ground flash densities in the tropical region (19 to 65 flashes.km^{-2}.year^{-1}) are higher than in the temperate region (17 to 27 flashes.km^{-2}.year^{-1}). Unfortunately, there is no LLS in the North part of Argentina (in the temperate region) and in the central part of Africa (in the tropical region), where the most intense thunderstorms on Earth in terms of total lightning activity occur (Zipser et al., 2006), to compare with the results in Tables 4.3 and 4.8.

Table 4.8. Absolute maximum ground flash density observed in different continents of the temperate region.

Continent	Period of Observation	Absolute Maximum Ground Flash Density[1] (flashes km^{-2} year^{-1})
North America	1996-2000	18
Europe	1992-2001	17
Asia	1997-2001	24
Africa	2006-2007	27

[1] Values were referred to for a spatial resolution of 1 km x 1 km.

Table 4.9 shows a comparison among peak current, multiplicity and percentage of positive ground flashes observed by Pinto et al. (2007) in Brazil, Orville and Huffines (2001) in the United States and Schulz et al. (2005) in Austria.

Table 4.9. Peak current of negative and positive ground flashes and the percentage of positive ground flashes observed by some LLS in the temperate region.

Country (Reference)	Positive Peak Current (kA)	Negative Peak Current (kA)	Percentage of Positive Ground Flashes (%)	Negative Multiplicity
Brazil (Pinto et al., 2007)	29.5	26.5	8.5	1.9
United States (Orville and Huffines, 2001)	23.5	24.5	4.5	2.4
Austria (Schulz et al., 2005)[2]	10.7	10.7	12.0	2.2[1]

[1] Only the 1999-2001 period was considered.

[2] Also a private communication.

An analysis of the values in Table 4.9 shows small differences between the values in Brazil and the United States, while the peak current values in Austria are considerably lower and the percentage of positive flashes considerably higher than the values in Brazil and the United States. If we consider that the LLS in Brazil and the United States (for the period corresponding to the observations of the Orville and Huffines, 2001) have very similar sensors (LPATS and IMPACT) and base lines (average of about 300 km), while the LLS in Austria has only IMPACT sensors installed at shorter base lines (approximately 100 km), the differences in the Austrian performance of the networks seem to reflect the differences of detection efficiency and cloud flash contamination. In turn, the differences between Brazil and the United States are very small and probably related to other LLS characteristics and not to geographical differences in the flashes.

Figures 4.14 to 4.17 show the normalized mean monthly distributions of the number of ground flashes, the percentage of positive ground flashes and the peak current of positive and negative ground flashes, following the methodology described by Pinto et al. (2007). All distributions are shifted so that the month numbered as seven corresponds to the month of largest activity. Except for the number of downward ground flashes, which is dependent on the area covered by the network, the absolute values of the other parameters were normalized (Pinto et al., 2007). In Figure 4.15, one can see that the period of largest lightning activity along the year in Brazil, defined arbitrarily as the number of months in which the lightning activity is 50% larger than the peak activity, is longer (six months) than in the United States and Austria (just three months). The mean monthly distribution of the percentage of positive flashes and positive peak current (Figures 4.16 and 4.17) are similar, while the mean monthly distribution of the peak current of negative flashes in Brazil (Figure 4.18) is different from those in the other countries. The difference is apparently related to a significant decrease in the mean negative peak current in Brazil during August/September in comparison with the other countries.

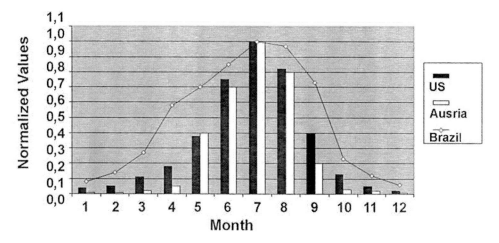

Figure 4.15. Normalized mean monthly distribution of the number of ground flashes observed in different countries for long time periods (adapted from Pinto et al., 2007).

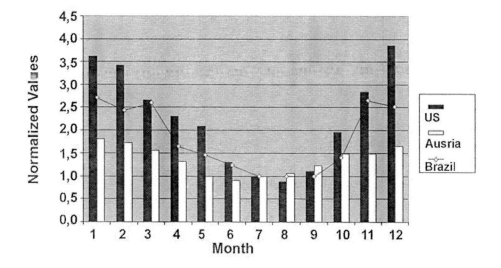

Figure 4.16. Normalized mean monthly distribution of the percentage of positive ground flashes observed in different countries for long time periods (adapted from Pinto et al., 2007).

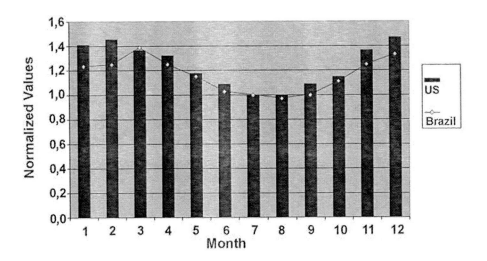

Figure 4.17. Normalized mean monthly distribution of the positive peak current of ground flashes observed in different countries for long time periods. Data for Austria are not available (adapted from Pinto et al., 2007).

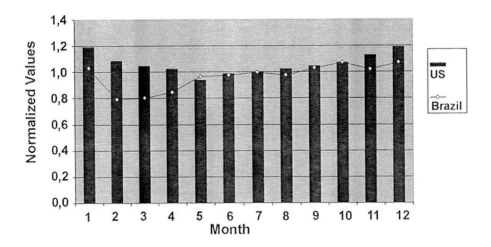

Figure 4.18. Normalized mean monthly distribution of the negative peak current of ground flashes observed in different countries for long time periods. Data for Austria are not available (adapted from Pinto et al., 2007).

Figure 4.19 shows the mean monthly distribution of the number of fires observed by satellite NOAA 12 during the same period (from 1999 to 2004) and same region of the data in Figure 4.18. Apparently, the decrease in negative peak current observed during August/September shown in Figure 4.18 is related to the injection of large amounts of smoke from fires during these months, since a significant correlation between the monthly distribution of number fire spots and the negative peak current was found. This result is consistent with the observations of Fernandes (2005), who found a decrease in mean monthly peak current of negative flashes from August to December in the North region of Brazil for two years of data and of Murray et al. (2000), who found a small decrease in the negative peak current of CG flashes in a thunderstorm event, both apparently related to the smoke from fires. Lyons et al. (1998a) found, for the same event of Murray et al. (2000), a large effect on the percentage and peak current of positive flashes. This effect was not found by Pinto et al. (2007).

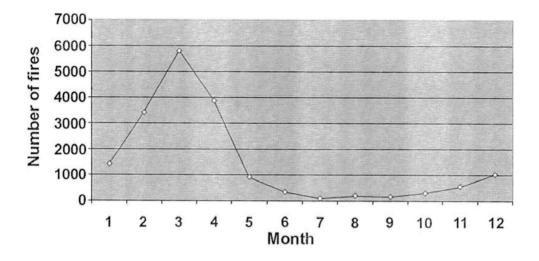

Figure 4.19. Mean monthly distribution of the number of fires for the same period and region in Figure 4.18 (adapted from Pinto et al., 2007).

Chapter 5

OBSERVATIONS BY OTHER TECHNIQUES

5.1. MAGNETIC AND ELECTRIC FIELD SINGLE DEVICES

Observations of lightning peak current by magnetic links in the tropical region were carried out in Brazil (Cherchiglia et al., 1998), Rhodesia (Anderson and Jenner, 1966), Peru (Foust et al., 1953) and the Java Island (Soetjipto et al., 1993). From these observations, the following median values for negative ground flashes were obtained: 46 kA (Brazil), 38 kA (Rhodesia), 42 kA (Peru) and 26 kA (Java Island). Frame aerials were used to record lightning peak current and multiplicity in the tropics by Lee et al. (1979). The observations were carried out in Kuala Lumpur in the 1970s. Median peak values for negative ground flashes of 36 kA and multiplicity of 4.9 were obtained. Torres et al. (1996) used a parallel-plate antenna and the stroke location obtained from a lightning location system to estimate a media peak current of 43 kA in Colombia. Torres et al. (2008) observed that all single device peak current measurements, including the observations by an instrumented tower in the Cachimbo Mountain station (see section 5.3), suggest that the peak current of negative downward ground flashes in the tropical region is higher than in the temperate region. However, most of these observations are not validated by independent calibrated instruments, so it is not possible to verify the influence of instrumental bias on them.

Observations of flash density by flash counters were extensively made from the 1950s to 1980s in many places around the world (Prentice, 1977; Mackerras and Darveniza, 1994; Rakov and Uman, 2003). Lightning flash counters are simple antenna-based instruments to record electric (or magnetic) field generated by lightning in general in the range from hundreds of hertz to tens of kilohertz. The most widely used counters are the so-called CIGRE 500-Hz and CIGRE 10-kHz counters (Anderson, 1977). Another largely used counter was the CGR1 and its modified version CGR3 (Mackerras, 1985; Mackerras and Darveniza, 1994). In the tropical region, observations were made by Sunoto (1985) and Soetjipto et al. (1989) in the Java Island, Romualdo et al. (1989) in Mexico, Mackerras (1978) and Kuleshov et al. (2006) in North Australia and Cherchiglia et al. (1998) in the southeast of Brazil.

Figure 5.1 shows a CIGRE 10 kHz flash counter installed in the state of Minas Gerais, in the Southeast Brazil, and a map of ground flash density obtained by a network of 43 of these counters obtained from 1985 to 1995 in the southeast of Brazil (Pinto et al., 2003a). The maximum flash density observed by this network was 14 flashes.km^{-2}.year^{-1}.

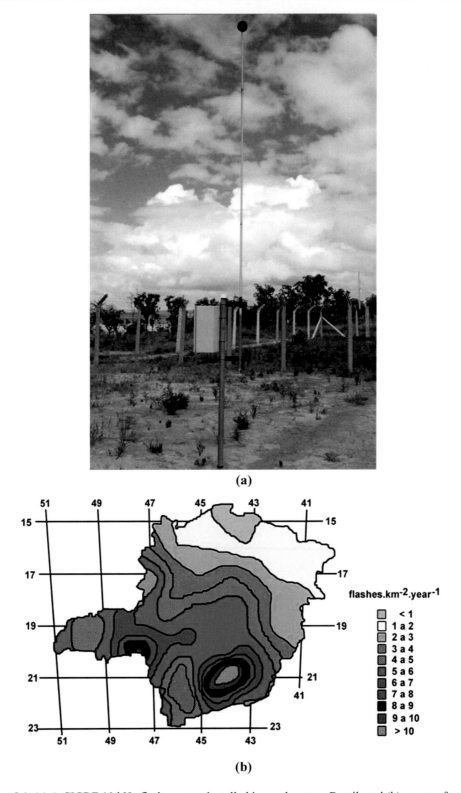

Figure 5.1. (a) A CIGRE 10 kHz flash counter installed in southeastern Brazil; and (b) a map of ground flash density obtained from a network of 43 of these counters from 1985 to 1994.

The maximum flash density reported by Kuleshov et al. (2006) in the north part of Australia with a CIGRE 500-Hz 13-counter network was 7 flashes.km^{-2}.year^{-1}. In Mexico, the maximum flash density reported by Romualdo et al. (1989) was 10 flashes.km^{-2}.year^{-1} in the south part of the country. In most cases, the maximum flash densities obtained by flash counters do not agree with more recent observations suggesting that they have a low reliability.

Filho et al. (2007) have used fast electric field antennas to study the electric field peak ratio between first and subsequent strokes in negative downward ground flashes in Brazil and compared their observations with similar ones made in other countries. They found that the ratio in Brazil is smaller than in Florida (Rakov et al., 1994) but higher than in Austria (Schulz and Diendorfer, 2006). Schulz et al. (2008) have extended the comparison to include observations in Sweden. They found that the ratio in Sweden is similar to that in Brazil.

5.2. HIGH-SPEED CAMERAS

The first lightning observations with high-speed cameras in the tropical region were made in southeast Brazil in 2000 using a digital video camera model Red Lake Motion Scope 8000S with a resolution and exposure time of one millisecond. All high-speed video recordings had a one second pre-trigger time and a total recording time of two seconds (2000 frames). The pre-trigger time of one second proved to be long enough to avoid missing the first strokes. Also, the total recording time of two seconds is long enough to capture the whole flash. All images were GPS synchronized, time stamped and without any image persistence (Saba et al., 2006a). Initially, the main goal of these observations was to validate the BrasilDat LLS data (Saba et al., 2004). Later, the observations were extended to investigate other lightning characteristics. The use of these cameras is particularly interesting to avoid bias introduced by finite video resolution of standard video tape recordings (Thomson et al., 1984). The advent of high-speed motion CCD video cameras (Figure 5.2) allowed the use of temporal high-resolution observation of lightning flashes. This observation technique can be considered an accurate-stroke-count technique, that is, a reliable way of studying lightning parameters such as multiplicity and percentage of single-stroke flashes that are sensitive to stroke counting (Saba et al., 2006a). With these cameras several processes (stepped leaders, return strokes, and continuing currents) that occur during a lightning flash can also be visualized with high temporal resolution and detail (see Chapter 2 for an illustration).

When making lightning observations with high-speed cameras, it is advisable to use a red filter in front of the lens in order to increase the contrast between the lightning channel and the background during diurnal recordings. It is also highly recommended that the video frames of the high-speed camera be GPS time-stamped to an accuracy of one millisecond. This synchronization allows the comparison of each flash recorded by the camera with the same flash detected by an LLS or other technique (e.g. electric-field measurements). In the data analysis of the high-speed observations, data from the LLS are used to obtain the stroke polarity, an estimate of the peak current near the ground, and the location of the ground strike point.

Figure 5.2. Photograph of the first high speed camera used in Brazil in 2000.

From 2003 to 2007, more than 1000 downward ground flashes were recorded with high-speed cameras in the Southeast region of Brazil during the summer season (Saba et al., 2006a, b; 2007a, b, c; 2008a, b, c; Pinto et al., 2008). The statistics of some parameters obtained from these observations for negative and positive downward ground flashes are summarized in Table 5.1 and Table 5.2, respectively.

Table 5.1. Parameters of negative downward ground flashes obtained by high-speed cameras in Brazil.

Parameter	Number of Events	Mean Value	Median Value
Negative leader speed (x 10^5 m s^{-1})	59	3.4	2.3
Percentage of single flashes (%)	233	20	-
Multiplicity	233	3.8	3.0
Flash duration (ms)	221	276	188
Number of strike points per flash	138	1.7	2.0
Interstroke interval (ms)	608	83	58
Continuing current duration (ms)	250	32	13.5

Table 5.2. Parameters of positive downward ground flashes obtained by high-speed cameras in Brazil.

Parameter	Number of Events	Mean Value	Median Value
Positive leader speed (x 10^5 m s^{-1})	10	2.6	1.9
Percentage of single flashes (%)	60	75	-
Multiplicity	60	1.3	1.0
Flash duration (ms)	53	175	142
Number of strike points per flash	56	1.2	1.0
Interstroke interval (ms)	18	190	154
Continuing current duration (ms)	65	104	68

The mean leader speed in Tables 5.1 and 5.2 are 2-D leader speeds based on the distances between the camera and the flashes and on the geometric characteristics of the camera and lenses used (Saba et al., 2008a, b, c). Although there are several published measurements of leader velocities for negative downward flashes, there is only one downward positive leader speed published by Berger and Vogelsanger (1966). Table 5.1 and Table 5.2 show that the speed of positive leaders is not very different from negative stepped leaders for downward ground flashes, as well as from spider leaders. Saba et al. (2008a, b, c) also report the existence of negative recoil leaders (RLs) that appear during the leader propagation of positive downward ground flashes. The RLs reveal branching structures that are not usually recorded when using conventional photography or video cameras.

The flash duration in Tables 5.1 and 5.2 is defined as the time interval between the occurrence of the first return stroke and the end of the continuing current (CC) following the last return stroke, if present. The values in these tables are similar to those reported by Berger et al. (1975), Diendorfer et al. (1998) and Visacro et al. (2004). It is interesting to note that although most of the positive flashes are single, the median value of the duration of positive flashes (142 ms) is similar to negative ones (188 ms). This similarity may be explained by the fact that nearly 75% of positive flashes contain at least one long CC, that is, a CC lasting more than 40 ms, while only 10% of negative flashes contain at least one long CC (Saraiva et al., 2008b). Another explanation is that interstroke intervals are longer in positive flashes. Saba et al. (2006a) also found that for a given number of strokes there is a minimum duration of negative flashes. Saraiva et al. (2008a) have extended the study of Saba et al. (2006a) including observations with high-speed camera in Tucson, Arizona, United States, and found that a maximum duration also seems to exist. Figure 5.3 from Saraiva et al. (2008a) show a scatter plot demonstrating the relation between the flash duration and the number of strokes per flash. Saraiva et al. (2008b) have also compared some characteristics of negative downward flashes observed with high-speed cameras in Brazil with measurements with the same cameras in Tucson, United States and found no significant differences.

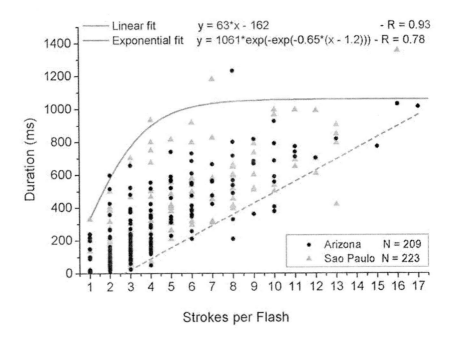

Figure 5.3. Scatter plot showing the relation between the flash duration and the number of strokes per flash. The black circles are data from Tucson and the gray triangles, from Brazil (adapted from Saraiva et al., 2008a, b).

Saba et al. (2006a) show that up to five different contact points were observed in a negative flash and the vast majority (90%) of the new terminations was created after there had been just one stroke in the previous channel. The average number of lightning strike points is 70% higher than the number of flashes. Considering that the missing of strokes in high-speed videos is practically negligible, it can be inferred from the average negative multiplicity (3.8) and from the average number of strike points per flash (1.7), that each ground contact point is, in average, struck 2.2 times. For positive downward flashes the average number of strike points per flash is 1.2 and all subsequent strokes in multiple-stroke flashes created a new termination on ground. In addition, Saba et al. (2006a) found that in negative downward flashes, long time intervals between strokes are usually associated with the presence of long CC. Less than 1% of all interstroke intervals observed last more than 500 ms, while about 20% of the interstroke intervals presented values lower than 33 ms. For positive downward ground flashes the geometric mean of the time intervals is about two times greater than the respective value for negative downward ground flashes. Sometimes a subsequent stroke in a downward positive flash occurs while the CC of the previous stroke is still occurring (never observed for negative downward ground flashes).

Saba et al. (2006a) also found that in negative downward flashes the presence of a long CC after first strokes is very rare. Only four first strokes of multiple-stroke flashes and six single flashes were observed followed by long CC. In addition, they found that on average, the longer the CC, the lower is the return stroke peak current preceding it.

Ballaroti et al. (2005) studied the presence of very-short CC (with duration between 3 and 10 ms) in negative downward flashes using high-speed cameras. From an analysis of 455 negative downward ground flashes, they found that 17% of all strokes presented very-short

CC, a higher occurrence if compared with short and long occurrences (11%). They also found that six strokes had two simultaneous ground contacts, indicating the occurrence of forked strokes.

Saba et al. (2006b) found from the observation of 454 negative strokes of downward flashes followed by CC with durations from 4 to 542 milliseconds that negative strokes with peak current higher than 20 kA are never followed by CC durations greater than 40 ms. On the other hand, negative strokes that have peak currents lower than 20 kA are followed by CC of any duration. Considering the high number of cases observed, these parameters determine an "exclusion zone" for negative strokes, which seems to indicate that high peak current negative strokes followed by long CC do not exist (Figure 5.4). As a consequence, they found that the detection efficiency of the BrasilDat LLS decreases from 62% for negative strokes followed by very-short CC to 36 % for strokes followed by long CC, with an intermediate value of 57% for strokes followed by short CC (from 10 to 40 ms).

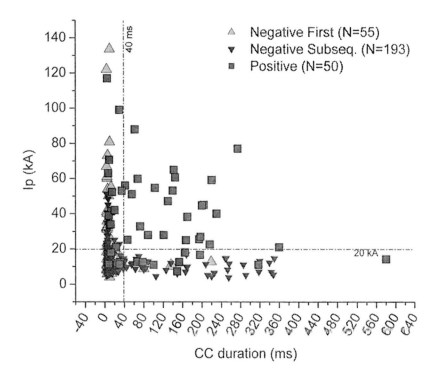

Figure 5.4. Peak current (I_p) versus CC duration for 248 negative strokes and 48 positive strokes of downward flashes (adapted from Saba et al., 2006b).

Saba et al. (2008c) and Campos et al. (2008a) have studied the occurrence of CC in positive downward flashes. They found that 74% of the positive flashes analyzed contained at least one long CC. They found that the "exclusion zone" found for negative strokes was not observed for positive strokes.

Campos et al. (2007) studied the time variations in the CC intensity for 63 negative downward ground flashes using high-speed cameras, considering that the luminosity of the channel is directly proportional to the current that flows through it. They can be either M-components (approximately symmetrical short duration current pulses superimposed to the current base level) or long duration variations that define the CC wave shape (Figure 5.5). Long duration variations can be grouped into six wave shape types. In extremely long CC more than 30 M-components have been observed. For the first time, evidence of M-components was observed in positive CG flashes, however only four of the six types observed for negative flashes were identified (Campos et al., 2008a). The average number of M-components in CC of positive flashes was 9, while for negative downward flashes it was 5.5. Campos et al. (2008b) have compared the waveforms of CC in Brazil with observations in Tucson, United States.

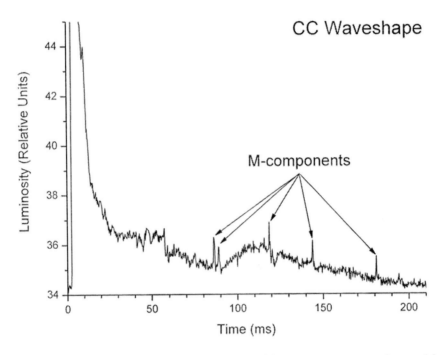

Figure 5.5. Example of a waveshape of continuing current with M-components superimposed for a negative downward ground flash. The M-components are indicated by arrows (adapted from Campos et al., 2007).

Ferro et al. (2008) analyzed the CC initiated by strokes following a new channel to ground in multiple stroke flashes using high-speed cameras, electric field measurements from a fast antenna and LLS data. They observed that the long continuing current initiated by a stroke that follows a new channel also obeys the pattern in the initiation of long continuing current suggested by Rakov and Uman (2003). We also verify that the suggestion given by Rakov and Uman (1990) that strokes initiating long-continuing currents tend to have lower initial electric field peaks than regular strokes is valid for strokes that create a new channel to ground and are followed by long CC. Apparently the reduction of peak current value when the stroke is followed by a long CC is stronger than the increase that is commonly observed when strokes follow a new channel. They also find that the "exclusion zone" proposed by Saba et al. (2006a) is valid for new channels initiating CC, and that the number of strokes in a same channel or the existence of a long CC current do not always consolidate the channel in a multiple stroke flash.

5.2. INSTRUMENTED TOWERS

Many instrumented towers have been used for lightning research in different countries. The most complete characterization return strokes of ground flashes, however, was made by instrumented towers in Switzerland from 1943 to 1971 (Berger, 1967, 1972, 1977; Berger et al., 1975). The current data obtained in Switzerland correspond to oscillograms measurements (above 2 kA) using resistive shunts installed at the top of two towers about 70-m high on the summit of Monte San Salvatore, Lugano. The summit of the mountain is 915 m above sea level. Because of the presence of the mountain, which contributes to enhance the electric field, the effective height of the towers was estimated to be 350 m (Eriksson, 1978a). For this reason, most ground flashes striking the towers were upward flashes. Berger et al. (1975) presented a summary of the waveform parameters of downward negative and positive flashes based on a sample size of 101 and 26 events, respectively. The following parameters were reported: peak current, charge, impulse charge (which excludes the CC), front duration, maximum dI/dt, stroke duration (defined as the time from the 2 kA threshold to the half-peak value on the tail), action integral and flash duration. In particular, the median negative and positive peak current values were 30 kA and 35 kA, respectively, and no negative peak current above 100 kA was observed.

However, Ballarotti et al. (2008) have claimed that since the data published by Berger et al. (1975) are from tower measurements and considering that about half of the natural downward flashes have more than one ground strike point (Rakov and Huffines, 2003; Rakov et al., 1994), there is a possible bias in some of Berger's flash parameters.

Direct measurements of ground flashes in instrumented towers in the temperate region has been made in many other countries such as Italy (Garbagnati et al., 1978; Garbagnati and Lo Piparo, 1982), Russia (Gorin et al., 1977; Gorin and Shkilev, 1984), South Africa (Eriksson, 1978b), Canada (Hussein et al., 1995, 2008; Janischewskyj et al., 1997), Germany (Beierl, 1992; Fuchs et al., 1998), Japan (Miyake et al., 1992; Goto and Narita, 1992, 1995), Switzerland (Montandon, 1992) and Austria (Diendorfer et al., 2000, 2008a, b).

In the tropical region, there is only one instrumented tower for lightning observations located in the Cachimbo Mountain station (CMS), geographic coordinates 43°58'W and 20°00'S, in Southeast Brazil. The tower was installed in the beginning of the 1980s and the observations began in 1985 (Trignelli et al., 1995). The tower in the CMS is very similar to the tower installed in South Africa some years earlier (Eriksson, 1979). In addition to the current of the flashes striking the 60m metallic tower, the CMS records the nearby ground lightning activity and the atmospheric electric field. It also provides photographic records and video images of the flashes striking the tower. The tower is located at the top of a mountain about 1430 m above sea level and 200 m above any other mountain in the region and has an effective height of 210 m. In South Africa the effective height is 148 m (Eriksson, 1978a). The current is measured by two current transducers (Rogowski coils), with a bandwidth from DC to 1 MHz, installed in the base of the tower. The accuracy of the current and sampling time of the data were initially limited to 760 A – 1 µs and 116 A – 0.2 µs for the main and the parallel transducer, respectively. The CMS uses a fiber optic link installed into an open duct with a copper plate ground system to transmit the information from the sensor to equipment room (Pinto et al., 2005, 2008). Figure 5.6 shows the tower and the current sensors located at its base and Figure 5.7 shows one example of a negative downward ground flash recorded in the CMS.

(a)

(b)

Figure 5.6. Photograph of (a) the tower in the CMS and (b) the current sensors located at its base (courtesy of Cemig Power Company).

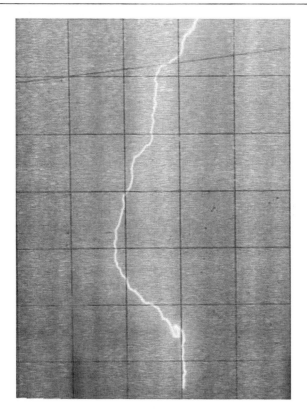

Figure 5.7. Photograph of a negative downward ground flash striking the CMS tower (courtesy of Cemig Power Company).

The first results obtained at CMS were published by Trignelli et al. (1995). Figures 5.8 and 5.9 show the current waveforms of the first three return strokes of a negative downward 5-stroke flash and the current waveforms of two strokes of a negative upward 3-stroke flash obtained at CMS, respectively.

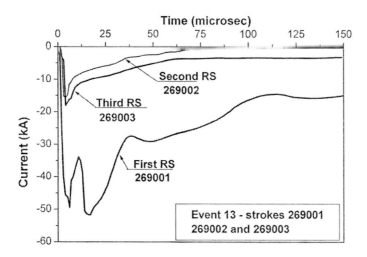

Figure 5.8. First three strokes of a negative downward 5-stroke multiple flash obtained at CMS.

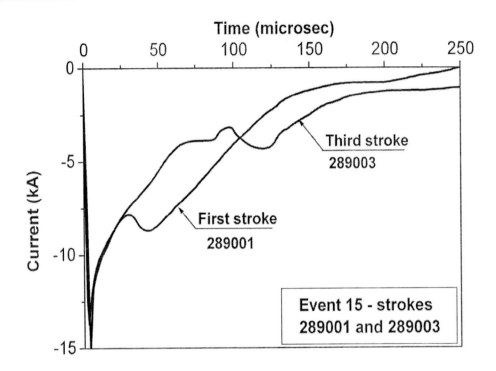

Figure 5.9. Negative upward flash with 3 return strokes obtained at CMS. The curve of the second stroke was not recorded.

A comparison among the observations in the CMS and the observations in Switzerland, Italy and South Africa (all outside the tropics) was first published by Pinto et al. (1997). They presented a comparative analysis of median peak current of first and subsequent return strokes of 27 negative ground flashes observed in CMS. Similar observations in Switzerland, Italy, and South Africa show that the values in the CMS are slightly higher than the value reported in South Africa and 30% higher than in Switzerland and Italy.

Lacerda et al. (1999) presented a waveform analysis of 22 negative downward lightning flashes obtained from 1985 to 1994 in CMS. In the analysis they computed the correlation between the instant when the peak current is maximum and the instant when the derivative current is maximum and identified M-components in subsequent strokes. In turn, Schroeder et al. (1999) and Mello et al. (2000) presented an analysis of the peak current of first strokes covering the period from 1995 to 1998.

Later, Pinto et al. (2003b), Visacro et al. (2004) and Pinto et al. (2005) extended the comparison to other waveform parameters. Pinto et al. (2003b) analyzed 30 downward flashes with a total of 88 return strokes captured in the CMS from 1985 to 2000, 29 with negative polarity and one with positive polarity. Only flashes with first stroke peak current higher than 2 kA were considered in the study. The downward flashes were identified on the basis of the early part of the first stroke current waveform and on photographs, when available. Visacro et al. (2004) summarize all waveform parameters for downward flashes based on a total of 33 flashes. Pinto et al. (2005) compared the peak current of first and subsequent strokes in the CMS with other similar data and triggered data.

Pinto et al. (2003b) also computed the percentage of downward and upward negative flashes from a total of 51 flashes captured at CMS from 1985 to 1998 and compared CMS observations with similar data obtained by the two towers in Switzerland (Berger et al., 1975) from 1963 to 1974, considering the results of the two towers separately. During this period, the towers in Switzerland (called here T1 and T2) captured 60 and 58 flashes, respectively. Considering that the towers are separated by only 400 m and located at almost the same height (the difference is about 47 m), the results of the analysis give an estimation of the intrinsic variability in the data set in Switzerland. Table 5.3 shows the results obtained by Pinto et al. (2003b).

Table 5.3. Lightning data at Switzerland (towers T1 and T2) and at CMS

Tower	Number of Events	Negatives Flashes				Positive Flashes	
			Downward				
		Upward	Total	Single	Multiple	Upward	Downward
T1	60	6 (10%)	42 (70%)	29 (69%)	13 (31%)	6 (10%)	6 (10%)
T2	58	2 (3.4%)	49 (84.5%)	32 (65.3%)	17 (34.7%)	2 (3.4%)	5 (8.7%)
CMS	51	10 (19.6%)	29 (56.9%)	14 (48.3%)	15 (51.7%)	11 (21.6%)	1 (1.9%)

Pinto et al. (2003b) concluded that the percentage of upward flashes with strokes more intense than 2 kA at CMS (41.2%) is much higher than those at the Switzerland towers (20% for T1 and 6.8% for T2), in spite of the similar height of the towers. This behaviour is the opposite of what would be expected if considering the effective height of the towers and its relation to the percentage of upward flashes (Rakov, 2003). The percentage of multiple negative downward flashes at CMS (51.7%) is higher than those in Switzerland (31% for T1 and 34.7% for T2), while the percentage of positive downward flashes at CMS (1.9%) is lower than those in Switzerland (10% for T1 and 8.7% for T2). More recently, this analysis was extended to consider changes in the first-stroke peak current in towers T1, T2 and CMS related to the time of the year. The data were divided in two groups: events in the summer season and events in the other seasons. The results indicate that while in towers T1 and T2 the peak current is significantly higher during summer than in other seasons (37 kA versus 29 kA), in the CMS the peak current remains the same. This result suggests that the peak current changes for different meteorological environments and, in consequence, in different regions.

Visacro et al. (2004) summarize all the waveform current parameters for downward negative ground flashes in the CMS, considering 33 flashes. Table 5.4 shows a comparison between these values and those reported by Berger et al. (1975) for Mount San Salvatore, Switzerland. They found that the peak currents in the CMS are higher than in San Salvatore, while the front duration and front steepness differ from those in San Salvatore by less than 20%. A detailed analysis of individual peak current values of first-stroke flashes in CMS and San Salvatore shows that the difference in the mean values is a result of the fact that no flashes with first-stroke peak currents below 20 kA were recorded in MCS. At the present time, there is no explanation for this feature. Visacro et al. (2004) also reported a percentage of single negative flashes of 48%, an average multiplicity of 3 and an average time between strokes of 64 ms.

Table 5.4. Negative downward ground lightning parameters obtained in the CMS (adapted from Visacro et al., 2004).

Waveform Current Parameter for Negative Downward Ground Flashes	Observed Value	
	CMS	Switzerland
Median First Stroke Peak current (kA)	45	30
Median Subsequent Stroke Peak current (kA)	16	12
Median First Stroke Front Duration T-10 and T-30 (µs)	5.6 and 2.9	4.5 and 2.3
Median Subsequent Front Duration T-10 and T-30 (µs)	0.7 and 0.4	0.6 and 0.4
Median First Stroke Front Steepness S-10 and S-30 (kA/ µs)	5.8 and 8.4	5.0 and 7.2
Median Subsequent Stroke Front Steepness S-10 and S-30 (kA/ µs)	18.7 and 24.7	15.4 and 20.1

Pinto et al. (2005) computed the peak current for first and subsequent return strokes of downward flashes and strokes of upward flashes in the CMS from 1985 to 2000, based on 29 downward and 6 upward flashes, with a total of 88 and 12 return strokes, respectively, and compared them with other similar observations. The results are shown in Table 5.5.

Table 5.5. Summary of the statistics on return stroke peak current of negative ground flashes in CMS and other countries (adapted from Pinto et al., 2005).

Type of Return Stroke	Location	Sample Size (strokes)	Median Peak Current[1] (kA)	Reference
First-stroke of Downward Flashes	Brazil	29	45	Pinto et al. (2005)
	Switzerland	101	30	Berger et al. (1975)
	South Africa	12	44	Eriksson (1979)
	Italy	42	33	Garbagnati and Lo Pipero (1982)
Subsequent strokes of Downward Flashes	Brazil	59	18	Pinto et al. (2005)
	Switzerland	135	12[2]	Berger et al. (1975)
	South Africa	8	18	Eriksson (1979)
	Italy	33	18	Garbagnati and Lo Pipero (1982)
Upward Flashes	Brazil	12	10	Pinto et al. (2005)
	Switzerland	176	10	Berger (1978)
	South Africa	1	10	Eriksson (1979)
	Italy	142	8	Garbagnati and Lo Pipero (1982)

[1]Only short towers are considered, since tall towers the current peaks are significantly influenced by the presence of the object (Rakov, 2001).
[2]Same value was reported by Anderson and Eriksson (1980) for a smaller sample size.

Pinto et al. (2005) concluded that the peak current of first-stroke of downward flashes are higher than all other strokes. They also concluded that the peak currents of subsequent strokes of downward flashes seem to be the same in all locations and the peak currents of strokes of upward flashes are lower than peak current of strokes of subsequent downward flashes in all locations, although they are initiated in a similar manner, that is, by downward dart leaders.

Figure 5.10 shows the current waveform of one of the two positive downward flashes measured at the CMS. The current waveform has a millisecond-scale and can be considered as belonging to the second category of waveforms of downward positive flashes, as described by Rakov and Uman (2003). Rakov and Uman (2003) argued based on Berger's observations that positive downward flashes can be divided into two categories. The first category includes microsecond-scale waveforms similar to those for negative downward flashes, while the second category includes millisecond waveforms with rise times up to hundreds of milliseconds. Examples of the two types of current waveform along with illustrations of the processes presumably leading to the formation of these waveforms are given in Rakov and Uman (2003).

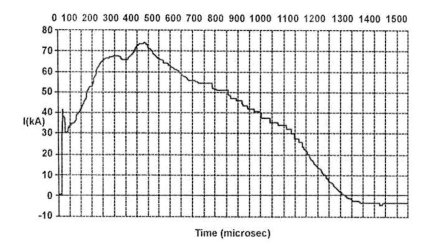

Figure 5.10. Current waveform of a downward positive ground flash recorded in the CMS.

The influence of instrumented towers on lightning parameters has been suggested by several authors (Risk, 1994; Borghetti et al., 2004), but for short-towers it is generally not confirmed by observations (CIGRE report, 1997; Diendorfer et al., 1998; Rakov and Uman, 2003; Guedes et al., 2003; Visacro and Silveira, 2005). In particular, Guedes et al. (2003) and Visacro and Silveira (2005) have evaluated the influence of the current sensor position along short-towers on the contamination of measured lightning current waves. The results have shown that, for realistic front time values of negative downward strokes, the current wave measured at tower top and at the base are practically the same, for both first and subsequent strokes.

5.3. ROCKET-AND-WIRE

This technique involves the launching of small rockets trailing thin ground (classical method) or ungrounded (altitude method) wires into the gap between the ground and a charged cloud overhead. The first triggered lightning was produced in 1960 in a research vessel off the west coast of Florida (Newman, 1965). The first triggered lightning produced over land was in France (Fieux et al., 1975). Triggered lightning experiments have been conducted in many countries (Rakov, 1999), including Japan (Nakamura et al., 1991, 1992), the United States (Rakov and Uman, 2003) and China (Liu and Zhang, 1998; Zhang et al., 2008).

Triggered lightning was observed in Brazil from 1999 to 2005 in the International Center for Triggered and Natural Lightning Studies, which was located in Cachoeira Paulista (22°41'S and 44°59'W; altitude of 625 m), a small city located halfway between São Paulo and Rio de Janeiro (Pinto et al., 2005). The triggering site was located on a flat 120m x 70m area on a hilltop (Figure 5.11). The control room was located 45 m away from the rocket launcher, which was able to launch up to 12 rockets during the same event. A field mill connected to the control room via fiber optic link monitored the ambient electric field. The plastic rockets are 0.85 m long and carry a wire spool. In addition to the current sensor (a 1-mW coaxial resistive shunt placed inside a metallic box) located under the rocket launcher (Figure 5.12). Several instruments were mounted around the launcher: standard video cameras, a high-speed digital camera, fast E-field sensors and optical sensors. The accuracy of the current measurements is considered to be better than 1 kA for currents below approximately 50 kA.

Figure 5.11. The launching system surrounded by three lightning rods and the control room behind it in the triggering site in Cachoeira Paulista, Brazil.

Figure 5.12.Detail of the current sensors in the triggering site in Cachoeira Paulista, Brazil.

Figure 5.13 shows the photograph of the first triggered flash obtained in the site in 1999 by the altitude method. The flash was triggered using a 30-m Kevlar insulator cable under the copper wire (Saba et al., 2005).

Figure 5.13. Photograph of the first triggered flash obtained in Brazil in 1999.

Pinto et al. (2005) summarized the current data obtained in Brazil, which correspond to peak currents of two flashes obtained by the classical method and one by the altitude method, with a total of 9 and 7 return strokes, respectively, and compared with similar observations in other countries. All strokes had peak currents above 2 kA. The results are shown in Table 5.6 for negative flashes. No positive flashes were recorded in the ICTNLS.

Table 5.6. Negative triggered flash parameters obtained in the ICTNLS (adapted from Pinto et al., 2005).

Method	Location	Number of Events	Peak Current (kA)	Reference
Classical[1]	Brazil	9	19	Pinto et al. (2005)
	Alabama, US	37	11	Fisher et al. (1993)
	Florida, US	305	12	Depasse (1994)
	France	54	10	Depasse (1994)
Altitude	Brazil	7	34	Pinto et al. (2005)
	Florida, US	2	21	Laroche et al. (1991) and Lalande et al. (1998)

[1] Other values reported in the literature can be found in Rakov et al. (1998) and Rakov (1999). There are no significant variations with respect to the values shown in this table.

Pinto et al. (2005) suggested that peak current of strokes of negative flashes obtained by the classical method may be the same in all locations. The variations reported in the literature can probably be explained by differences in the number of events in the data sets and/or by different time intervals chosen to launch a rocket. However, seasonal effects cannot be disregarded. Also, the peak current values of strokes of negative flashes obtained by the classical method are similar to those of strokes of upward flashes, and less intense than subsequent strokes of downward natural flashes. A possible explanation for why strokes of negative flashes obtained by the classical method are less intense than subsequent strokes of downward negative ground flashes is that they are induced at times when no conditions to produce a subsequent stroke of a natural ground flash are reached. There are only nine direct measurements and three indirect estimates of the peak current of strokes of negative flashes obtained by the altitude method.

5.4. VLF MEASUREMENTS

Two techniques have been used to measure lightning activity in the VLF range: VLF LLS and Schumann resonance observations in single stations. LLS in this frequency range can cover large areas since the signal can circle the globe without much attenuation by repeatedly bouncing off the conducting ionosphere and the Earth. At the present time, three LLS cover the tropical region: the World Wide Lightning Location Network - WWLLN (Lay et al., 2004; Jacobson et al., 2006), the ATDnet (Gaffard et al., 2008) and the ZEUS Long-Range Monitoring Network (Anagnostou et al., 2002; Rodriguez et al., 2007).

The World-Wide Lightning Location Network (WWLLN) is the result of an international collaboration among research institutions across the globe. The current network has 25 VLF (3-30 kHz) sensors typically 7000 km apart. Each lightning stroke location requires the time of group arrival (TOGA) from a least 4 WWLLN sensors. The aim of the WWLLN is to provide global real-time locations of ground lightning, with reasonable detection efficiency and mean location accuracy. To achieve these values, the number of sensors and their geographical arrangement is important. A uniform spacing of sensors around Earth is ideal. Around 50 sensors would be needed to cover the whole world uniformly, with sensors spaced about 3000 km apart in order to achieve the above values in all regions. In South America, the first sensor was installed in São José dos Campos (Brazil) in the beginning of 2004 and the first analysis of the WWLLN performance was made by Lay et al. (2004). They found detection efficiency in South America of the order of a few percents. In 2005, two other sensors were installed in Huancayo (Peru) and Cordoba (Argentina) which increased the detection efficiency in South America to about 10%. Figure 5.14 shows a map of flash density in Brazil obtained from the WWLLN during the period from 2006 to 2007 (Naccarato and Pinto, 2008).

The ATDnet has been successfully operated by the UK Met Office for nearly 20 years. The original range includes all of Europe. Recent expansions and improvements to the network have increased the range of detectable lightning to include all of South America, Africa and central Asia since December 2007 (Gaffard et al., 2008).

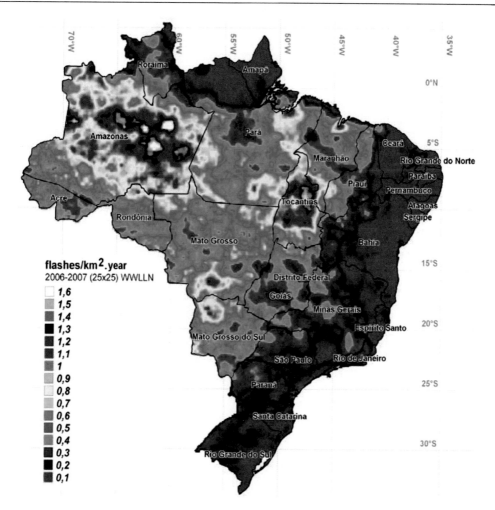

flashes/km^2.year
2006-2007 (25x25) WWLLN

- 1,6
- 1,5
- 1,4
- 1,3
- 1,2
- 1,1
- 1
- 0,9
- 0,8
- 0,7
- 0,6
- 0,5
- 0,4
- 0,3
- 0,2
- 0,1

Figure 5.14. Map of lightning activity in Brazil obtained by the WWLLN from 2006 to 2007 (adapted from Naccarato and Pinto, 2008)

The improved network currently has 14 detectors, operating in the frequency of 13.7 kHz, all of them located in Europe and the North African continent, outside the tropics. While in central Europe the network has a detection efficiency around 505 and a location accuracy of 5-6 km (Gaffard et al., 2008), no validation studies have been reported for the tropical region.

The ZEUS system consists of a network of thirteen VLF receivers measuring radio noise emitted by lightning in the 7-15 kHz frequency range, most of them located in the African continent. A preliminary validation of the network shows that the location accuracy ZEUS is between 10-50 km over Brazil, 5-25 km in Africa, 70-100 km over the southern EUA, and 5-25 km in the Atlantic Ocean (Rodriguez et al., 2007). The detection efficiency of the network seems to vary from 10 to 40% in the African continent (Chronis et al., 2007), being lower in other regions.

Schumann resonance signals can be seen as the AC component of the global electric circuit, resulting from the integration of the global lightning activity. This is achieved by the ELF electromagnetic waves produced by lightning being trapped in the natural Earth-ionosphere waveguide, reflecting off the ionosphere and Earth's surface. The size of the waveguide is such that resonances occur close to 8, 14, 20, … Hz, known as the Schumann resonances (Schumann, 1952). Observations by Schumann resonance (SR) methods in single stations have been made since 1960s (Heckman et al., 1998; Nikolaenko and Rabinowscz, 1995).

The SR observations have shown that in the Earth-ionosphere cavity, lightning activity is concentrated mainly in three major areas: Southeast Asia, Africa and South America. This is in agreement with the satellite lightning observations. The magnitude of the lightning activity in these three regions reaches its peak at different times in the diurnal cycle (Sentman, 1995). Sentman and Fraser (1991) and Price and Melnikov (2004) have shown the diurnal variation of the global lightning activity in these three major lightning regions. The total lightning activity in Southeast Asia, Africa, and South America reaches maximum at approximately 08:00 UT, 14:00 UT, and 22:00 UT, respectively. The largest global lightning activity occurs during the northern hemisphere summer. This is also in agreement with satellite lightning observations. Moreover, Sentman and Fraser (1991) showed that the inter-annual variability of global lightning activity differs for each of the three major source regions of global lightning. In September, the lightning activity in South America is stronger than that in Africa. In April, the lightning activity increases by 40% in Africa, and decreases by 20% in South America. The lightning activity in South-east Asia remains approximately the same during these two months. Heckman et al. (1998) using SR observations suggested a global flash rate of 22 flashes.$km^{-2}.s^{-1}$.

Chapter 6

LIGHTNING AND GLOBAL WARMING

Lightning incidence and distribution is directly linked to the Earth's climate, which is basically driven by solar heat and light, that is, total solar irradiance. On the diurnal and seasonal scales, lightning activity peaks in most regions (including the tropical region) a few hours or months after the peak solar heating. Globally, due to the asymmetry between the hemispheres, with more land areas in the northern hemisphere, more lightning occurs in that area during summer in relation to the southern hemisphere summer. However, recent findings cast doubt that the variation of the solar irradiance can explain the climate changes in course, since the observed variation seems too small to affect global warming (Foukal, 2003). Most evidence suggests that human activity is probably responsible for global warming, at least at the rate at which it has been observed (IPCC report, 2007).

In this scenery, predicting lightning changes in response to global warming is very complex, because it encompasses many different factors that respond differently when the atmosphere warms up. These factors include changes in regional temperature, atmospheric dynamics, precipitation, ocean circulation and frequency and amplitude of El Niño/La Niña phenomena. The term El Niño was initially used to describe a warm-water current that periodically flows along the coast of Ecuador and Peru, disrupting the local fishery. This oceanic event is associated with the fluctuation of a global-scale tropical and subtropical surface pressure pattern called the Southern Oscillation. This coupled atmosphere-ocean phenomenon, with preferred time scales of two to about seven years, is collectively known as the El Niño-Southern Oscillation (ENSO). It is often measured by the surface pressure anomaly difference between Darwin and Tahiti and the sea surface temperatures in the central and eastern equatorial Pacific. During an ENSO event, the prevailing trade winds weaken, reducing upwelling and altering ocean currents such that the sea surface temperatures warm, further weakening the trade winds. This event has a great impact on the wind, sea surface temperature and precipitation patterns in the tropical Pacific. It has climatic effects throughout the Pacific region and in many other parts of the world, through global teleconnections. The cold phase of ENSO is called La Niña. The linkage between global warming and lightning changes is further complicated by the fact that each of these processes may have different response time-scales.

From the observational point of view, the reasons for not finding long-term trends in the global lightning activity, if they exist, may be due to the short time period with available data for large regions (less than two decades) and/or partly due to the fact that during this time period the change in temperature was less than 0.5 °C, which may indicate a low sensitivity of the global lightning activity to global warming. The low sensitivity could be attributed to the difference between the spatial distribution of the global lightning activity, which indicates that lightning is on average considerably more frequent in the tropics than in the temperate and polar regions, and the region more affected by the global warming, which is located at high latitudes. However, considering the nonlinearity of the climate system, it is possible that small temperature changes may lead to abrupt changes in the lightning activity.

The lack of known long-term trends in the lightning activity may explain why there is no specific reference to future changes in lightning activity in the last (Forth) Assessment Report of the Intergovernmental Panel on Climatic Changes (IPCC) report (IPCC report, 2007). In fact, the IPCC report indicates that there is insufficient evidence to determine whether trends exist in such events as tornadoes, hail, lightning and dust storms, which occur in small spatial scales.

From the modeling point of view, although models have evolved over time, one of the main problems in predicting the response of lightning activity to global warming is the still limited spatial resolution of the present General (or Global) Climate Models (GCMs). GCMs, like other weather models, use known laws of physics with prescribed initial and boundary conditions of the atmosphere to compute its evolution through time. Although based on physical laws, they have different ways of dealing with processes that cannot be explicitly represented based on physical laws. For instance, the smallest features that the present GCMs can resolve are hundreds of km wide; in consequence, physical processes that take place on a much smaller scale than the models, like thunderstorms, cannot be resolved. In order to improve the spatial resolution of the present GCMs it is necessary to better understand the complexes coupling and feedback mechanisms involving different parts of the models. An interaction mechanism between processes is called a feedback mechanism when the result of an initial process triggers changes in a second process that in turn influences the initial one. A positive feedback intensifies the original process, and a negative feedback reduces it. For instance, it is necessary to understand how the tropical Pacific will respond to increasing temperatures. The last 50 years observations suggest that there is no evidence of trends in the variability or the persistence of the ENSO (Nicholls, 2008). In addition, GCM results present diverging views of the Pacific response to global warming (Vecchi et al., 2008).

6.1. CLIMATE CHANGES AND GLOBAL WARMING

Climate in a narrow sense is usually defined as the average weather, or more rigorously, as the statistical description in terms of the mean and variability of relevant quantities over a period of time ranging from months to thousands or millions of years. The climate has changed throughout every period of the Earth's history. In the Holocene Epoch, a warm period that began 10,000 years ago and corresponds to the Human Era, the temperature alternated warmer and cooler millennial-scale periods of a few degrees. About 6000 years ago, the average temperature was apparently 2° C higher and there was greater rainfall than today. At that time, the Sahara was far more fertile than today supporting large herds of animals. Then, the climate started to cool up to the end of the 16th century, when an almost constant cold period of about 100 years known as "little ice age" occurred, especially in

Europe. After this age, the climate changed again and started to warm. Since 1900, the global climate has warmed by about 0.8°, 0.6° of which in the last 30 years. Figure 6.1 shows the global average surface temperature anomaly from 1860 to 2000, with respect to the 1961 to 1990 mean.

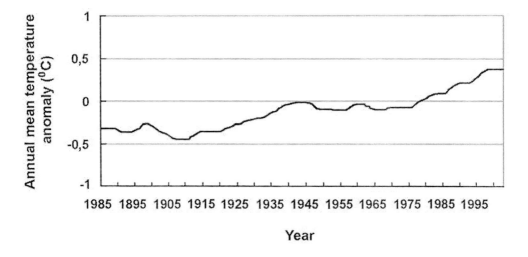

Figure 6.1. Global annual mean surface temperature anomaly from 1985 to 2003, with respect to the 1961 to 1990 mean (adapted from Brohan et al., 2006).

The temperatures in the polar and temperate regions in the northern hemisphere are rising much more quickly than in the southern hemisphere and in the tropics. Since 1900, the temperature in the polar and temperate regions of the northern hemisphere has increased by about 1.2°C, while the temperature in the polar and temperate regions of the southern hemisphere and in the tropics has increased by about 0.7 °C.

Differently from the past temperature changes that are attributed to natural internal processes or external solar forcings, the present change has been shown to be associated with the rising of greenhouse gases due to anthropogenic activities, rather than other natural causes (IPCC report, 2007). This is because, differently from oxygen and nitrogen, the predominant gases that compose the Earth's atmosphere, which absorb virtually no infrared radiation and allow it to escape to space, greenhouse gases absorb and reradiate the infrared radiation to the Earth's surface (which ultimately also allows life to exist). Since the temperature of Earth is controlled by the balance of the input of energy of the sun (mainly visible and ultra violet radiation) and its loss back into space, if the concentration of greenhouse gases increases, the global temperature increases as well. The main greenhouse gases are carbon dioxide (CO_2), methane (CH_4), nitrous oxide (N_2O) and chlorofluorocarbons (CFCs). Figure 6.2 shows the annual mean concentration of CO_2 in the atmosphere from 1959 to 2000 in the Mauna Loa Mountain in Hawaii, a location far from local sources of pollution. It shows that atmospheric concentrations of carbon dioxide have increased every single year since 1958 and the increase is believed to be mainly due to human sources such as fossil fuel combustion, including industrial process, and land use changes, including deforestation (IPCC report, 2007). Figure 6.3 shows the large deforestation in the Amazon region in Brazil produced by the cut down of forests for agriculture, releasing large amounts of carbon dioxide into the atmosphere. Moreover, the removal of trees reduces the ability to remove subsequent inputs of carbon dioxide by photosynthesis. In fact, only little more than half of the increase in net radiation accounted for by the gases cited above is due to carbon dioxide.

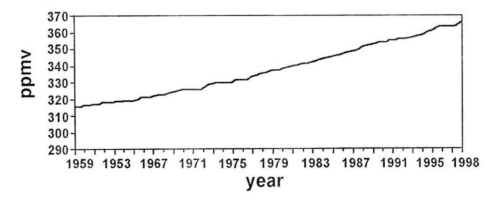

Figure 6.2. Annual mean concentration of carbon dioxide in the atmosphere (in ppmv) from 1958 to 2000 in the Mauna Loa mountain in Hawaii.

Figure 6.3. Deforestation in the Amazon region in Brazil.

In the beginning of this century, evidence of the increase in the Atlantic Ocean hurricane activity (Webster et al., 2005), in severe thunderstorms with more powerful tornadoes (Trapp et al., 2007; Marsh et al., 2007), in severe thunderstorms over the tropical oceans (Aumann et al., 2008), in rainfall extremes (Allan and Soden, 2008; Rajeevan et al., 2008), and in the upper-tropospheric humidity particularly in the tropics (Bates and Jackson, 2001) has been reported. In general they have been associated with the global warming. In addition, evidence from GCMs has predicted an increase in the global rainfall (IPCC report, 2007; Del Genio et al., 2007), although it should decrease in some regions or even in much of the tropics (IPCC report, 2007).

The Forth Assessment Report of IPCC presents the results of recent GCMs that predict a range of possible outcomes (known as climate scenarios) for different levels of elevated atmospheric carbon dioxide concentration (IPCC report, 2007). The model predictions indicate a "best scenario" increase in the global average temperature of about 1 °C in the year 2100, relative to the 2000 value, and a "worst scenario" of as much as 6 °C. Most model predictions indicate that the smallest temperature changes will occur in tropical latitudes. In this region, annual changes on the order of 0.1 °C to 3 °C are predicted, depending on model assumptions and location.

6.2. LIGHTNING CONTRIBUTION TO GLOBAL WARMING

The atmosphere of Earth is composed by about 21% of molecular oxygen (O_2), about 78% of molecular nitrogen (N_2), together with a number of trace gases, such as argon, helium and radiatively active greenhouse gases such as carbon dioxide and ozone. In addition, the atmosphere contains water vapour, whose amounts are highly variable, clouds and aerosols.

As lightning occurs, it produces new trace molecules from the ambient constituents of the atmosphere, such as nitrogen dioxide (NO_2) and nitrogen monoxide (NO), which together are often referred to as NO_x. The nitric oxide (NO) is the most important because it facilitates chemical reactions in the troposphere and stratosphere that determines the concentrations of ozone (O_3).

Although in the upper troposphere NO_x is also produced by other processes, lightning is believed to be the dominant source, mainly in the tropical region (MacGorman and Rust, 1998; Singh et al., 1999). In the upper troposphere, ozone is a greenhouse gas. In consequence, lightning-induced NO_x in the upper troposphere has important implications for the climate (Schumann and Huntrieser, 2007; IPCC report, 2007), since perturbations in the lightning may have a large effect on ozone (Toumi et al., 1996; Thompson et al., 2000; Martin et al., 2002). Mickley et al. (2001) found that observed long-term trends in ozone over the past century might be explainable by an increase in lightning.

Lightning also produces greenhouse gases by biomass burning in large wildfires or bushfires. Biomass burning in the tropics and at high latitudes is likely to increase with climate change, both as a result of increased lightning and as a result of increasing temperatures and dryness (Price and Rind, 1994; Stocks et al., 1998; Williams et al., 2001; Brown et al., 2004). It is believed that lightning causes more than a hundred thousand wildfires worldwide each year, mainly in drought areas like California, in the United States, or parts of Australia, launching tons of carbon oxides and other greenhouse gases into the atmosphere. In the United States alone, about 30,000 wildfires are triggered by lightning. In the western United States, there is evidence indicating that they have been more frequent and destructive during the last 20 years when compared with the previous 20 years. This situation has been attributed to the rising average temperature in the region which produces a longer fire season (Uman, 2008).

If the wildfires burn for some time, they may also produce vigorous rising air currents and release large quantities of water vapor to the air which may affect thunderstorm electrification (Lyons et al., 1998a; Fernandes et al., 2006) and even form a pyro-thundercloud.

The impact of lightning on the production of greenhouse gases may cause a positive feedback mechanism (Williams, 1992), in such a way that an increase in the lightning activity in response to increasing temperature causes more greenhouse gases. They in turn increase the surface air temperature by radiative forcing causing more lightning. However, other aspects may influence this mechanism. Recent results have shown a connection between tropical lightning activity, ice water content (Petersen et al., 2005) and ice crystal size (Sherwood et al., 2006), and all of these may affect the radiative forcing.

Several GCMs identify an increase in lightning-induced NO_x due to global warming (Schumann and Huntrieser, 2007). The estimates of increases in NO_x vary within 4 to 60% for 1 °C of temperature change, with median near 15%. Some studies found no global trend in lightning-induced NO_x over the period of 1990–2030, but found significant changes in its distribution (Stevenson et al., 2005; Sanderson et al., 2006). The precise knowledge of the lightning-induced NO_x is important to understand and predict the role of the feedback mechanism between climate changes and lightning.

6.3. Lightning Response to Global Warming

Even though lightning activity is a result of nonlinear microphysical and thermodynamical processes acting through the entire troposphere and affected by many meteorological variables, it is well established that it is sensitive to variations in the surface air temperature at many temporal scales (Williams, 1992, 1994, 1999, 2005, 2008; Price, 1993; Markson and Price, 1999; Reeve and Toumi, 1999; Price and Asfur, 2006a; Sekiguchi et al., 2006; Markson, 2007; Pinto and Pinto, 2008a). The well known universal time variation of the global electrical circuit is a global manifestation of that temperature sensitivity (Price, 1993). Williams (1992) showed that monthly mean SR magnetic field observed in Kingston, Rhode Island during a 5.5-yer period over one strong El Niño cycle is quite correlated with the monthly mean tropical surface air temperature. This correlation suggests that the Schumann resonance amplitude can be used as a global tropical thermometer. The longest available record on Schumann resonance intensity shows a slight decline with time (Williams, 2008; Satori et al., 2008). This negative slope is however not statistically significant.

In terms of temporal scales, lightning is generally more prevalent on average in the afternoon than at night and in the summer than in the winter, in direct response to variation in the solar insolation (Price, 2008; Pinto and Pinto, 2008a). It is also generally expected that global lightning activity tends to increase in the climate scale in response to global warming (Williams, 1992; Price and Rind, 1994; Brasseur et al., 2005; Hauglustaine et al., 2005). However, some authors did not find long-term trends in the lightning activity using different techniques (Price and Asfur, 2006b; Stevenson et al., 2006; Markson, 2007; Harrison, 2006), although many recent studies indicate a high positive correlation between surface air temperature and lightning activity (Williams, 2005; Price and Asfur, 2006a; Sekiguchi et al.,

2006; Pinto and Pinto, 2008a). While long-term changes in the mean lightning activity have been not found, almost no studies have been made addressing possible trends on extreme lightning activity events. Such events can potentially produce severe impacts on our society and environment (Nicholls and Alexander, 2007).

In a global perspective, lightning sensitivity to changes in global surface air temperature has been estimated from the analysis of lightning data obtained by satellites (Reeve and Toumi, 1999; Ming et al., 2005; Petersen and Buechler, 2007), lightning-related thunderstorm day records (Changnon, 1988, 2001), lightning-related upper-tropospheric water vapor (Price and Asfur, 2006a), lightning-related global atmospheric circuit ionospheric potential (Markson, 2007; Williams, 2007) and lightning-dominated SR intensity changes (Williams, 1992; Sekiguchi et al., 2006). However, the observational analyses are limited by the still short periods, of the order of one decade (satellite lightning data and Schumann resonance data); the low accuracy of data over an extended period (most of thunderstorm days data and upper-tropospheric water vapor data) and to the weaker sensitivity of the data to lightning changes (ionospheric potential data).

The first estimates of lightning sensitivity to increasing temperature based on satellite data was made by Reeve and Toumi (1999) based on OTD lightning data for three years. They found that the variations in global monthly land lightning activity are well correlated with variations in global monthly land surface wet bulb temperature and estimated a 40% increase in the global lightning activity for 1 °C of temperature change. The correlation was strongest in the northern hemisphere, weak in the southern hemisphere and absent in the tropics. Ming et al. (2005) extended the work of Reeve and Toumi (1999) using 5 years of OTD data and 8 years of LIS data. They found that on the inter-annual time scale the global flash rate increases by 17% for 1 °C of temperature change. Over land in the northern hemisphere the increase was 13%, while no trends were found in the southern hemisphere and in the tropics. Later estimates using satellite data, however, do not confirm this result (Petersen and Buechler, 2007). Petersen and Buechler (2007) failed to find any clear trend in the global lightning activity with increasing temperature in the OTD and LIS satellite data for the decade of 1998-2007, although regionally some trends were observed.

One of the difficulties to observe trends in lightning activity with increasing temperature at annual time scale (Pinto and Pinto, 2008a) is related to the strong influence of El Niño/La Niña over lightning activity, mainly in the tropics (Nickolaenko and Rabinowicz, 1995; Heckman et al., 1998; Satori et al., 2007, 2008). The situation is still more complex because the impact of global warming on the intensity or frequency of these phenomena is poorly understood (Cobb et al., 2003).

If on the one hand no trends are observed in experimental data, on the other hand most GCMs predict an increase in the lightning activity (Schumann and Huntrieser, 2007). Price and Rind (1994) used a GCM and assumed a double concentration of carbon dioxide to estimate the lightning sensitivity to increasing global surface air temperature. They used a simple parameterization that assumes cloud top height as a proxy for lightning activity and two different relations for continental and maritime thunderstorms (Price and Rind, 1992). They found a lightning sensitivity of approximately 6% for 1 °C of temperature change.

Michalon et al. (1999) extended the work by Price and Rind (1994) including both cloud top height and droplet concentration in the parameterization, trying to take into account the role of the microphysical cloud characteristics. They found a lower increase in the surface air temperature for a double carbon dioxide condition, but with almost the same lightning sensitivity to increasing temperature found by Price and Rind (1994).

Later, it was realized that the cloud top height does not primarily control the formation of lightning (Molinié and Pontikis, 1995; Price et al., 1997; Ushio et al., 2001; Allen and Pickering, 2002; Cecil et al., 2005). The cloud top heights may be large without active updrafts and hence without active lightning. More recently, many other approaches were developed based on other parameters (Schumann and Huntrieser, 2007). Grenfell et al. (2003) and Shindell et al. (2006) based also on GCMs have shown an approximate 10% increase in the lightning activity for 1 °C increasing temperature with most of the increase in the tropics.

On the other hand, most of the model results presented in the Forth Assessment Report of the IPCC (IPCC report, 2007) suggest that the greatest warming due to increasing greenhouse gases will occur in the upper tropical troposphere and not on the surface, due to the enhanced water vapor injected in this region by deep convection. As a consequence, the atmosphere would become more stable, which could in principle reduce the lightning activity in contrast with the results indicated by Grenfell et al. (2003) and Shindell et al. (2006).

The paradox above has been interpreted by Del Genio et al. (2007) as indicative that the number of thunderstorms in a warmer climate will be lower but more energetic, producing more lightning. However, it may also be interpreted as a consequence of the many limitations in the present GCMs. Among these limitations, the GCMs are not able to consider the high spatial resolution to resolve convective activity, the atmosphere-ocean coupling, particularly important considering that most of the tropical region is formed by oceans, and the aerosol effects on convection and photochemical-related processes. In addition, the GCMs also need to include detailed surface emissions and height-dependent meteorological fields. In particular, there is a lack of more complete information about the NOx production by different sources and tropospheric meteorological data, mainly in the tropical region. Such limitations cause a considerable variation in the predictions of such models, therefore no definitive conclusions concerning the future impact of global temperature changes on the lightning activity can be reached.

The lightning sensitivity to changes in global surface air temperature can also be investigated based on physical arguments (Williams, 2005). According to Williams (2005), a weaker sensitivity of global lightning is expected in longer time scales in response to a convective adjustment. This theory is based on the assumption that for larger time scales the increase in the surface air temperature can be communicated to the temperature aloft, causing a lesser change in the convective available potential energy (CAPE), which is the vertical integral of parcel buoyancy between the level of free convection and the level of neutral buoyancy. The idea is supported by the observations by Gettelman et al. (2002) and the model results were obtained by Ye et al. (1998). Gettelman et al. (2002) observed a small positive trend in CAPE in the tropics during 1958-1997, based on observations from 15 radiosonde stations. Ye et al. (1998) have used a GCM to investigate the possibility of predicting climate changes in CAPE based on its variability in the current climate, and on the observation of a linear relationship between CAPE and the surface wet-bulb potential temperature during the Australian Monsoon Experiment by Williams and Renno (1993). They found an increase in CAPE in response to a climate warming, although the climate sensitivity of CAPE to surface temperature is an order of magnitude smaller than the observed sensitivity to variations in the current climate. However, lightning occurrence is a very nonlinear function of convection intensity (Solomon and Baker, 1994), so that it is difficult to conclude what will be the impact on lightning activity produced by the climate changes in CAPE. Also, the global response of CAPE to climate warming is an integration of local responses that cannot be predicted by the present GCMs.

In a local perspective, in turn, the temporal variations of the lightning activity can also be very complex, mainly in short time scales (less than one decade), since local aspects may be as important as global aspects. Many local aspects could be important, such as changes in the local circulation of air masses in response to teleconnections related to El Niño processes and changes related to urban effects in association with heat islands (related to the nature of the surface and their response to solar radiation) and pollution (Steiger et al., 2002; Naccarato et al., 2003; Pinto et al., 2004; Pinto and Pinto, 2008a). The effect of these aspects on the lightning activity is illustrated in Figure 6.4 and Figure 6.5. Figure 6.4 shows the variation of the number of flashes recorded in the Southeast region of Brazil from 1999 to 2004, indicating a large increase in 2000-2001 related to La Niña event. In turn, Figure 6.5 shows a map of the mean annual lightning density in the region of the city of São Paulo for a spatial resolution of 1 km x 1 km during the period from 1999 to 2006, showing a large increase in adjacent areas (Pinto and Pinto, 2008a). The regions in white in Figure 6.5 represent values of flash density larger than 11 flashes.km^{-2}.year^{-1}, while regions in purple represent values of flash density lower than 3.5 flashes.km^{-2}.year^{-1}.

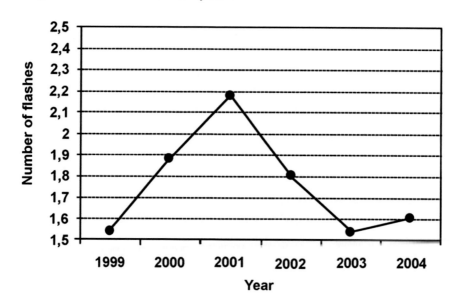

Figure 6.4. Variation of the number of flashes recorded in the Southeast region of Brazil from 1999 to 2004, indicating a large increase in 2000-2001 related to La Niña event (adapted from Naccarato, 2005).

In order to study the sensitivity of lightning to increasing temperature in long time scales, the only parameter related to lightning incidence for which worldwide data are available is thunderstorm days. This parameter has been shown to be related to the ground flash density through a power law relationship (Prentice, 1977; Anderson et al., 1984; Rakov and Uman, 2003).

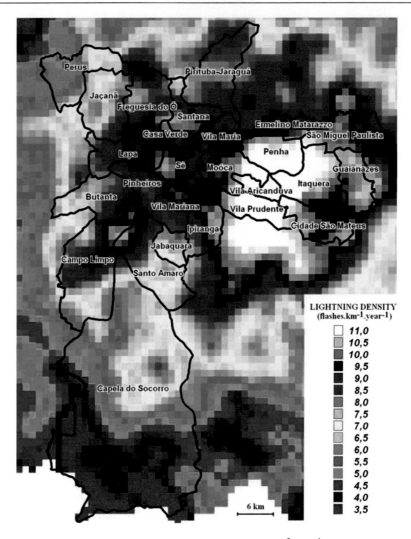

Figure 6.5. Map of the mean annual lightning density (in flashes.km^{-2}.year^{-1}) in the city of São Paulo for a spatial resolution of 1 km x 1 km for the period from 1999 to 2006. The regions in white represent values of flash density larger than 11 flashes.km^{-2}.year^{-1}, while regions in purple represent values of flash density lower than 3.5 flashes.km^{-2}.year^{-1} (adapted from Pinto and Pinto, 2008a).

An attempt to use thunderstorm days to study the lightning sensitivity to increasing temperature in long time scales was made by Pinto and Pinto (2008a). They used ground lightning data collected by the Brazilian Lightning Detection Network (BrasilDat) from 1999 to 2006 and thunderstorm day data from 1951 to 2006 to investigate the lightning sensitivity to changes in surface air temperature in the city of São Paulo, Brazil, for different time scales. The environment of large cities has been observed to suffer a dramatic impact of human activity on the local weather. The weather in large cities is appreciably different from the surrounding countryside. Buildings and roads absorb and store more sunlight and industrial processes and automobiles increase the pollution and create heat. In consequence, the climatic changes in these regions are in general larger in magnitude than the climatic changes associated with the global warming.

Pinto and Pinto (2008a) consider the city of São Paulo in their study for two reasons: first, the very high lightning activity in the city (a large area with flash densities above 11 flashes.km^{-2}.year^{-1} and a peak flash density of 17 flashes.km^{-2}.year^{-1}) and, second, the observed and documented increase in the surface air temperature in the last decades (Cabral and Funari, 1997). Both aspects are apparently related to the existence and growth of the large urban area of São Paulo in recent decades and the heat island effect associated with it (Naccarato, 2003; Farias et al., 2008). However, they may also be partially related to the global warming. Figure 6.6 shows the thunderstorm day and temperature data in the city of São Paulo from 1951 to 2006 (Pinto and Pinto, 2008a).

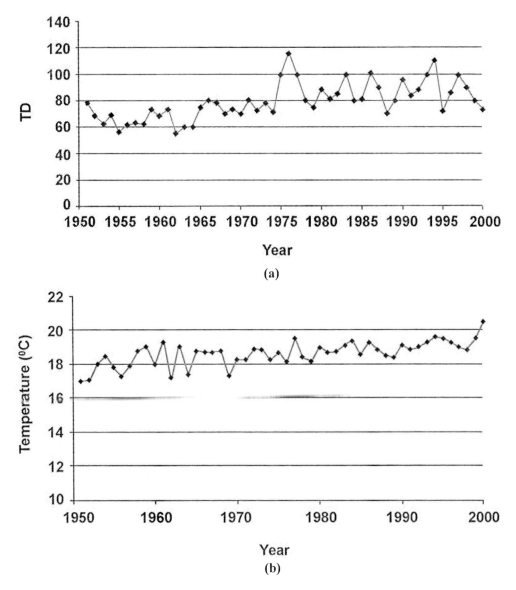

Figure 6.6. Data of (a) thunderstorm days and (b) temperature in the city of São Paulo from 1951 to 2000 (adapted from Pinto and Pinto, 2008a).

Figure 6.7 shows the decadal thunderstorm day values and the decadal mean surface air temperatures from 1950 to 1990 in the city of São Paulo. A reasonable agreement between both variables can be observed. In order to estimate the decadal increase in the number of flashes per 1 °C of temperature, Pinto and Pinto (2008a) converted the thunderstorm days into number of flashes using the power law relationship obtained comparing thunderstorm day and lightning data in the city during the last decade. Figure 6.8 shows the relationship between the decadal mean numbers of flashes (divided by 10000) and the decadal mean surface air temperatures in the city of São Paulo from 1950 to 1990. A linear fit and its squared correlation coefficient are indicated, as well as the standard deviations associated with the mean decadal numbers of flashes. The correlation tested by the nonparametric Spearman test is significant. From the linear fit, the mean increase in the number of flashes with temperature is 30% per 1 °C of temperature. This value is close to the 45% increase per 1 °C of temperature in the number of thunderstorms in the tropical ocean reported by Aumann et al. (2008).

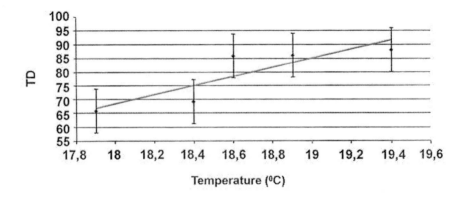

Figure 6.7. Relationship between the decadal thunderstorm day values and mean surface air temperatures for the decades of 1950 to 1990 in the city of São Paulo (adapted from Pinto and Pinto, 2008a).

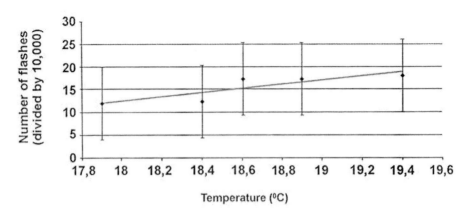

Figure 6.8. Relationship between the decadal mean numbers of flashes (divided by 10000) and mean surface air temperatures in the city of São Paulo for the decades 1950 to 1990. A linear fit and its squared correlation coefficient are indicated, as well as the standard deviations associated with the decadal mean numbers of flashes (adapted from Pinto and Pinto, 2008a).

The lightning sensitivity for the decadal time scale is lower than for daily and monthly scales, 40% per 1 °C of temperature (Pinto and Pinto, 2008a), suggesting that the lightning sensitivity to surface air temperature decreases for longer time scales in the city of São Paulo. This finding is in agreement with what is expected based on convective adjustment (Williams, 2005). Pinto and Pinto (2008b) have analyzed thunderstorm and temperature data for Hong Kong, in China, based on observations made by Kai-Hing (2003) and found results similar to those for São Paulo.

Pinto and Pinto (2008b) have also compared normalized thunderstorm data for three cities for the last 25 years: São Paulo, in Brazil, Hong Kong, in China, and Odessa, in the United States, and found very similar variations throughout this period (Figure 6.9). This result suggests that global and local lightning activities are related. However, even though the variations seem to be related to ENSO, the similarity is not quite as evident as would be expected, suggesting that other large scale phenomena may also have a significant influence on local lightning activity. A better understanding of the links between large scale and local phenomena related to lightning activity is crucial to predict the response of global lightning activity to increasing temperature.

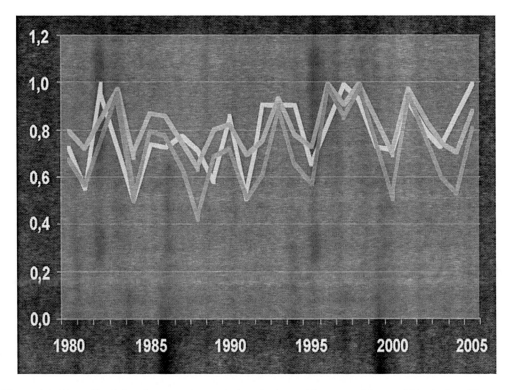

Figure 6.9. Normalized thunderstorm day variations in three cities around the world in the last 25 years: São Paulo, in Brazil (curve in light gray), Hong Kong, in China (curve in white), and Odessa, in the United States - curve in dark gray (adapted from Pinto and Pinto, 2008b).

Chapter 7

FINAL REMARKS

Besides reviewing for the first time the lightning literature in the tropical region, the goal of this book is contribute to two key questions that have been investigated for many years: Are ground lightning characteristics in tropical regions different from the other regions on Earth? How will lightning be in the future in a warmer planet?

Answering the first question, for the first time the amount of information about ground lightning in the tropical region is at a point where this question can be addressed with reasonable reliability. Ground lightning data obtained in Brazil in the last two decades by a lightning location system, high-speed cameras and an instrumented tower, complemented with data obtained by other techniques, are the largest and most reliable data set ever obtained in the tropical region. This fact gives a unique opportunity to characterize ground flashes in the tropics. The comparison of ground lightning characteristics in Brazil with data in the other countries in the tropics or in the temperate region, however, is incomplete due to the limitations in the techniques used. Some of these limitations are the low stroke DE of LLS, the absence of peak current information on high-speed camera data, the low spatial coverage of flash counters and the influence of the local orography on the observations by instrumented towers. Many other aspects should also be taken into account when comparing lightning data obtained by different techniques, among them: sensor calibration of LLS sensors, sensitivity of high-speed cameras to low luminosity features, the inability of instrumented towers to record strokes striking different points on ground and the physical differences of natural and triggered lightning. To summarize, while the ground flash density is clearly higher in the tropical region than in the temperate region, there is not sufficient data to prove that the peak current or any other characteristic of ground flashes in the tropical region are different from the temperate region.

For the second question, the most probable answer is yes, but with considerably greater uncertainty. The observations over the last two decades indicate that lightning clearly responds to temperature. If the results obtained in large urban areas and by most of the present general climate models, as well as the available evidence that more lightning is present in the warm El Nino phase than the cold La Nina phase in the global tropics, are used as surrogate the answer is an increase in the lightning activity and, probably, a change in the lightning patterns around the globe. From the physical point of view, however, the answer to this question hangs on the critical issue of 'convective adjustment': how the global temperature profile follows the surface air temperature on longer (climate) time scales. The problem is that we are not sure exactly what that adjustment is. If CAPE is invariant with warming, one might not have much response of global lightning to temperature. However, CAPE is a delicate quantity and we have little information on its long-term behavior, at least over a period comparable to the global record of surface air temperature. In addition, the possible role of an increase in lightning activity on Earth's climate, accelerating the rate at which the temperature changes remains to be investigated.

In conclusion, tropical lightning may have again other practical application for human beings as a monitoring system to climatic changes in response to the global warming. Also, it may be useful to monitor or even predict extreme atmospheric events such as hurricanes or severe storms. In a near future, lightning should be observed continuously from space by a geostationary satellite, as it already is observed on ground in many countries, making it an easier and more useful tool for monitoring climate changes.

REFERENCES

Abidin, H.Z.; Ibrahim, R. *Proceedings of the Nuclear Power and Energy Conference (PECon), 2003*, Bangi, Malaysia.

Abdullah, N.; Yahaya, M.P. *Proceedings of the 29th International Conference on Lightning Protection, 2008*, Uppsala, Sweden.

Allan, R.P.; Soden, B. *J. Science,* 2008, doi: 10.1126/science.1160787.

Allen, D. J; Pickering, K. E. *J. Geophys. Res.*, 2002, doi:10.1029/2002JD002066.

Anagnostou, E.N.; Chronis, T.; Lalas, D.P. *EOS-Transactions*, 2002, 83, 589, 594-595.

Anderson, R.B. *Physics of Lightning,* 1977, vol. 1, pp. 437-463, New York, Academic Press.

Anderson, R.B.; Jenner, R.D. *Trans. South Africa Inst. Elect. Engrs.*, 1966.

Anderson, R. B.; Eriksson, A. *J. Electra*, 1980, 69, 65-102.

Anderson, R.B.; Eriksson, A.J.; Kroninger, H.; Meal, D.V.; Smith, M.A. Lightning and Power Systems, 1984, London, *IEE Conf. Publ.* No. 236.

Aumann, H.H.; Ruzmaikin, A.; Teixeira, J. *Geophys. Res. Lett.*, 2008, 35, doi:10.1029/2008GL034562.

Ballarotti, M.G.; Saba, M.M.F., Pinto Jr., O. *Geophys. Res. Lett.*, 2005, 32, doi:10.1029/2005GL023889.

Ballarotti, M.G.; Saba, M.M.F.; Pinto Jr., O. *Proceedings of the 19th International Lightning Detection Conference (ILDC), 2006,* Tucson, AZ.

Ballarotti, M.G.; Saba; M.M.F., Pinto Jr., O.; Schulz, W. J. Geophys. Res., 2008, submitted.

Bates, J.J.; Jackson, D.L. *Geophys. Res. Lett.*, 2001, 28, 1695-1698.

Beierl, O. *Proceedings of the 21st International Conference on Lightning Protection (ICLP), 1992,* Berlin, German.

Berger, G.; Zoro, R. *Proceedings of the 1st International Conference on Lightning Physics and Effects (ICLP), 2004,* Belo Horizonte, Brazil.

Berger, K. J. *Franklin Institute*, 1967, 283, 478-525.

Berger, K. Bull. Schweiz. *Elektrotech,* 1972, 63: 1403-1422.

Berger, K. *Physics of Lightning,* 1977, Ed. R.H. Golde, vol. 1, Academic Press, New York, pp. 119-190.

Berger, K. Bull. Schweiz. *Elektrotech,* 1978, 69, 353-360.

Berger, K.; Vogelsanger, E. *Planetary Electrodynamics,* 1969, Eds. S.C. Coroniti and J. Hughes, New York, Gordon and Breach, pp. 489-510.

Berger, K.; Anderson, R.B.; Kroninger, H. *Electra,* 1975, 80, 223-237.

Bernardi, M.; Pigini, A.; Diendorfer, G.; Schulz, W., *CIGRE Report* , 2002, 33-205.

Bhikha, B.; Ojelede, M.E.; Annegarn, H.J.; Kneen, M. *Proceedings of the LIS International Workshop, 2006*, Huntsville, AL.

Biagi, C.J.; Cummins, K.L.; Kehoe, K.E.; Krider, E.P. *J. Geophys. Res.*, 2007, 12, doi:1029/2006JD007341.

Blakeslee, R.J.; Bailey, J.C.; Pinto Jr., O.; Athayde, A.; Renno, N.; Weidman, C.D. *Proceedings of the XII International Conference on Atmospheric Electricity (ICAE), 2003*, Versailles, France.

Boccippio, D.J.; Goodman, S. J. *J. Appl. Met.*, 2000, 39, 2231-2248.

Boccippio, D.J.; Koshak, W.; Blakeslee, R.; Driscoll, K.; Mach, D.; Buechler, D.; Boeck, W.; Christian, H.J.; Goodman, S.J. *J. Atmos. Oceanic Technology*, 2000, 17, 441–458.

Boccippio, D.J.; Cummins, K. L.; Christian, H. J.; Goodman, S. J. *Mon. Wea. Rev.*, 2001, 129, 108-122.

Boeck, W. L.; Suszcynsky, D. M.; Light, T. E., Jacobson, A. R.; Christian, H. J.; Goodman, S. J., Buechler, D. E., Guillen, J.L.L. *J. Geophys. Res.*, 2004, doi:10.1029/ 2003JD004491.

Bond, D.W.; Steiger, S. R.; Zhang, X. T.; Orville, R.E. *Atmos. Environment*, 2002, 36, 1509-1519.

Borghetti, A.; Nucci, C. A.; Paolone, M. *IEEE Transactions on Power Delivery, 2004*, 19, 03-09.

Brasseur G.P.; Schultz, M.; Granier, C.; Saunois, M. T.; Botzet, M.; Roeckner, E.; Walters, S. *J. Climate,* 2005, 19, 3932-3951.

Brohan, P.; Kennedy, J.J.; Haris, I.; Tett, S.F.B.; Jones P.D. *J. Geophys. Res.* 2006, 111, doi:10.1029/2005JD006548.

Brooks, *C.E.P. Geophys. Mem.* 1925, 24, 147–64.

Brown, T.J.; Hall, B.L.; Westerling, A.L. *Clim. Change*, 2004, 62, 365-388.

Burrows, W.R.; King, P.; Lewis, P.J.; Kochtubajda, B.; Snyder, B.; Turcotte, V. *Atmos. Ocean.*, 2002, 40, 59-81.

Cabral, E.; Funari, F. L. *Bull. Clim. FCT/UNESP*, 1997, 2, 218-222 (in Portuguese).

Campos, D.R.; Pinto Jr., O. *Proceedings of IX International Symposium on Lightning Protection (SIPDA), 2007*, Foz do Iguaçu, Brazil.

Campos, L.Z.S.; Saba, M.M.F.; Pinto Jr., O.; Ballarotti, M.G. *Atmos. Res*, 2007, 84:302-310.

Campos, L.Z.S.;, Saba, M.M.F.;, Pinto Jr., O.; Ballarotti, M.G. *Atmos. Res.*, 2008a, doi:10.1016/2008.02.020.

Campos, L. Z. S.; Saba, M. M. F.; Cummins, K. L.; Pinto Jr., O.; Krider, E. P.; Fleenor, S. A. *Proceedings of the International Lightning Detection Conference, 2008*b Tucson, AZ.

Cecil, D. J.; Goodman, S. J.; Boccippio, D. J.; Zipser, E. J.; Nesbitt, S. W. *Mon. Wea. Rev.,* 2005, 133, 543-566.

Cecil, D.J.; Zipser, E.J.; Liu, C.; Nesbitt, S.W. *Proceedings of the LIS International Workshop, 2006*, MSFC, Huntsville, AL.

Changnon Jr., S.A. *J. Climate,* 1988, 1, 389-398.

Changnon Jr., S.A. *J. Applied Met.*, 2001, 40, 783-794.

Chen, S.M.; Du, Y.; Fan, L.M.; He, H.M., Zhong, D.Z. *IEEE Trans. Electromagn. Compat.,* 2002, 44, 555-560.

Chen, S.M.; Du, Y.; Fan, L.M. *IEEE Trans. Power Delivery*, 2004, 19, 1148-1153.

Chen, A. B.; Kuo, C.-L.; Lee, Y.-J.; Su, H.-T.; Hsu, R.-R.; Chern, J.-L.; Frey, H. U.; Mende, S. B.; Takahashi, Y.; Fukunishi, H.; Chang, Y.-S.; Liu, T.-Y.; Lee, L.-C. *J. Geophys. Res.*, 2008, doi:10.1029/2008JA013101.

Chisholm, W.A.; Cummins, K.L. *Proceedings of the LIS International Workshop, 2006*, Marshall Space Flight Center, Huntsville, AL.

Cherchiglia, L.C.L.; Carvalho, A.M.; Diniz, J.H.; Souza, V.J. *Proceedings of the International Conference on Grounding and Earthing (GROUND'98), 1998*, Belo Horizonte, Brazil.

Christian, H.J. *Proceedings of the American Meteorological Society Meeting, 2007*, New Orleans.

Christian, H.J.; Blakeslee, R.J.; Goodman, S.J. *J. Geophys. Res.* 1989, 94: 13,329-13,337.

Christian, H.J.; Blakeslee, R. J.; Boccippio, D. J.; Boeck, W.L.; Buechler, D.E.; Driscoll, K.T.; Goodman, S. J.; Hall, J. M.; Koshak, W. J.; Mach, D.M.; Stewart, M. F. *J. Geophys. Res.*, 2003, doi:10.1029/2002JD002347.

Chronis, T.; Williams, E.; Anagnostou, E.; Petersen, W. *EOS-Transactions, 2007, 88*, 397-398.

Cianos, N.; Oetzel, G.N.; Pierce, E.T. *J. Appl. Meteor.*, 1973, 12, 1421-1423.

CIGRE report, Document 118, 1997.

Cobb, K.M.; Charles, C.D.; Cheng, H.; Edwards, R.L. *Nature*, 2003, 424, 271-276.

Cramer, J. A.; Cummins, K. L.; Morris, A.; Smith, R.; Turner, T. R. *Proceedings of the 18th International Lightning Detection Conference (ILDC), 2004*, Helsinki, Finland.

Cummins, K.L.; Bardo, E.A. *Proceedings of the International Conference on Lightning Physics and Effects (LPE), 2004*, Belo Horizonte, Brazil.

Cummins, K. L.; Hiscox W. L.; Pifer A. E.; Maier, M. W. *Proceedings of the 9th International Conference on Atmospheric Electricity (ICAE), 1992*, St. Petersburg, Russia.

Cummins, K. L.; Burnett, R. O.; Hiscox, W. L.; Pifer, A. E. *Proceedings of the Precise Measurements in Power Conference, 1993*, National Science Foundation and Center for Power Engineering at Virginia Tech, Arlington.

Cummins, K. L.; Bardo, E. A.; Hiscox, W. L.; Pyle, R. B.; Pifer, A. E. *Proceedings of the International Aerospace & Ground Conference on Lightning and Static Electricity, 1995*, National Interagency Coordination Group, Williamsburg.

Cummins, K. L.; Krider, E. P.; Malone, M. D. *IEEE Trans. Electromagnetic Compatibility,* 1998a, 40, 465-480.

Cummins, K. L.; Murphy, M. J.; Bardo, E. A.; Hiscox, W. L.; Pyle, R. B.; Pifer, A. E. *J. Geophys. Res.*, 1998b, 103, 9035-9044.

Del Genio, A.D.; Mao-Sung, Y.; Jonas, *J. Geophys. Res. Lett.*, 2007, 34, doi:10.1029,/2007GL030525.

Diendorfer, G. *Proceedings of the Symposium on Lightning Protection (SIPDA), 2007*, Foz do Iguaçu, Brazil.

Diendorfer, G. *Proceedings of the 29th International Conference on Lightning Protection, 2008b*, Uppsala, Sweden.

Diendorfer, G.; Schulz, W.; Rakov, V. A. *IEEE Trans. Electromagnetic Compatibility,* 1998, 40, 452-464.

Diendorfer, G.; Mair, M.; Schulz, W.; Hadrian, W. *Proceedings of the 25th International Conference on Lightning Protection (ICLP), 2000*, Rhodes, Greece, pp. 44-47.

Diendorfer, G.; Bernardi, M.; Cummins, K.; De La Rosa, F.; Hermoso, B.; Hussein, A.; Kawamura, T.; Rachidi, R.; Rakov, V.; Schulz, W.; Torres, H. *CIGRE report,* 2008a.

Diendorfer, G.; Cummins, K.; Rakov, V.A.; Hussein, A.M.; Heidler, F.; Mair, M.; Nag, A.; Pichler, H.; Schulz, W.; Jerauld, J.; Janischewskyj, W. *Proceedings of the 29th International Conference on Lightning Protection, 2008*b, Uppsala, Sweden.

Dwyer, J.R.; Rassoul, H.K.; Al-Dayed, M.; Caraway, L.; Chrest, A.; Wright, B.; Kozak, E.; Jerauld, J.; Uman, M. A.; Rakov, V. A.; Jordan, D. M.; Rambo, K. *J. Geophys. Res. Lett.,* 2005, doi:10.1029/2004GL021782.

Eriksson, A.J. Trans. South African IEE, 1978a, 69: 2-16.

Eriksson, A.J. *CSIR Special Report Elek,* 1978b, 152, National Electrical Engineering Research Institute, Pretoria, South Africa.

Eriksson, A. J. Ph. D. *Thesis, 1979, University of Natal, Pretoria, South Africa.*

Evert, R.; Schulze, G. *Proceedings of the IEEE PES Conference and Exposition in Africa, 2005,* Durban, South Africa.

Falcón, N.; Pitter, W.; Muñoz, A.; Nader, D. *IngUC,* 2001, 7, 47-53.

Falcon, N.; Quintero, A.; Ramirez, l. *Proceedings of the International Conference on Atmospheric electricity (ICAE), 2007,* Beijing, China.

Farias, W.R. G.; Pinto Jr., O.; Naccarato, K. P.; Pinto, I.R.C.A. *Atmos. Res.,* 2008, in press.

Fernandes, W.A., PhD. *Thesis, 2005, Brazilian Institute of Space Research - INPE,* 161 p. (in Portuguese).

Fernandes, W. A.; Pinto, I. R. C. A.; Pinto Jr., O.; Longo, K. M.; Freitas, S. R. *Geophys. Res. Lett.,* 2006, doi:10.1029/2006GL027744.

Ferro, M.A.S.; Saba, M.M.F.; Pinto Jr., O. Atmos. Res., 2008, in press.

Fieux, R.; Gary, C.; Hubert, P. *Nature,* 1975, 257, 212-214.

Filho, A.O.; Schulz, W.; Saba, M.M.F.; Pinto Jr., O.; Ballarotti, M. Proceedings of the International Conference on Atmospheric Electricity (ICAE), 2007, Beijing, China.

Fisher, R. J.; Schnetzer, G. H.; Thottappillil, R.; Rakov, V. A.; Uman, M. A.; Goldberg, J. D. *J. Geophys. Res.,* 1993, 98, 22887-22902.

Foukal, P. *EOS Transactions,* 2003, 84, 205-208.

Foust, C.M.; Maine, B.C.; Lee, C. *AIEE Trans. Power Appa.Syst.,* 1953, 72, 383-393.

Fuchs, F.; Landers, E.U.; Schmidt, R.; Wiesinger, J. *IEEE Trans. Electromagn. Compat.,* 1998, 40: 444-451.

Füllekrug, M.; Price, C. *Meteorol. Z.,* 2002, 11, 99-103.

Gaffard, C.; Nash, J.; Atkinson, N.; Bennett, A.; Callaghan, G.; Hibbett1, E.; Taylor, P.; Turp, M.; Schulz, W. *Proceedings of the International Lightning Detection Conference (ILDC), 2008,* Tucson, AZ.

Galvin, J.F.P., *Weather,* 2008, 63: 31-36.

Garbagnati, E.; Lo Piparo, G.B. *ETZ-A, 1982,* 103, 61-65.

Garbagnati, E.; Giudice, E.; Lo Piparo, G.B. *ETZ-A, 1978,* 99: 664-668.

Gettelman, A.; Seidel, D.J.; Wheeler, M.C.; Ross, R.J. *J. Geophys. Res.,* 2002, 107, 4606-4611.

Gill, T., *Proceedings of the International Lightning Detection Conference (ILDC), 2008,* Tucson, AZ.

Gorbatenko, V.; Ershova, T.; Rybina, N.; Thern, S., *Proceedings of the 29th International Conference on Lightning Protection (ICLP), 2008,* Uppsala, Sweden.

Gorin, B.N.; Shkilev, A.V. *Elektr.,* 1984, 8, 64-65 (in Russian).

Gorin, B.N.; Levitov, V.I.; Shkilev, A.V. *Elektrichestvo,* 1977, 8, 19-23 (in Russian).

Goto, Y.; Narita, K. *Res. Lett. Atmos. Electr.*, 1992, 12, 57-60.

Goto, Y.; Narita, K.I. *J. Atmos. Terr. Phys.*, 1995, 57, 449-458.

Grenfell, J.L.; Suindell, D.T.; Grewe, V. Atmos. Chem. Phys. Discuss., 2003, 3, 1805-1842.

Guedes, D. G. ; Pinto Jr., O.; Visacro Filho, S., *Proceedings of the VI International Symposium on Lightning Protection (SIPDA), 2003*, Curitiba.

Hagenguth, J.H.; Anderson, J.G. *AIEE Trans.,* 1952, 71, 641-649.

Harrison, H. *Geophys. Res. Lett.*, 2006, 33, doi:10.1029/2006GL025880.

Hauglustaine, D.A.; Lathière, J.; Szopa, S.; Folberth, G. *Geophys. Res. Lett.,* 2005, 32 doi:10.1029/2005GL024031.

Heckman, S.; Williams, E.; Boldi, R. J. Geophys. Res., 1998, 103, 31775-31779.

Heidler, F.; Zischank, W.; Wiesinger, J. *Proceedings of the 25th International Conference on Lightning Protection, 2000,* Rhodes, Greece.

Herodotou, N.; Chisholm, W. A.; Janischewskyj, *W. IEEE transactions on Power delivery,* 1993, 8, Nb. 3.

Hidayat, S.; Ishii, M. *J. Geophys. Res.*, 1998, 103, 14001-14009.

Hidayat, S.,; Ishii, M. *J. Geophys. Res.*, 1999, 104, 24449-24454.

Hidayat, S.; Ishii, M.; Hojo, J.; Sirait, K.T.; Pakpahan, P. *Proceedings of the 10th International Conference on Atmospheric Electricity, 1996,* Osaka, Japan.

Holle, R.L.; Lopez, R.E. Proceedings of the International Conference on Lightning and Static Electricity, 2003, Blackpool, England, Royal Aeronautical Society, 103-134, KMS, 7pp.

Horner, F. *Proceedings of IEEE*, 1954, 101, 383-390.

Hussein, A.M.; Janischewskyj, W.; Chang, J-S.; Shostak, V.; Chisholm, W.A.A.; Dzurevych, P.; Kawasaki, Z-I. *J. Geophys. Res.*, 1995, 100, 8853-8861.

Hussein, A.M.; Todorovski, D.; Milewski, M.; Cummins, K.L.; Janischewskyj, W. *Proceedings of the 29th International Conference on Lightning Protection (ICLP), 2008,* Uppsala, Sweden.

Ianoz, M. *IEEE Transactions on Electromagnetic Compatibility*, 2007, 49, 224 – 236.

Idone, V.P.; Saljoughy, A.B.; Henderson, R.W.; Moore, P. K.; Pyle, R.B. *J. Geophys. Res.,* 1993, 98, 18323–18332.

IPCC report, Intergovernmental Panel on Climate Changes (IPCC), Climate change 2007: The physical Science Basis, 2007, World Meteorological Organization (WMO) und UN Environment Program (UNEP).

Jacobson, A.R.; Knox, S.O.; Franz, R.; Enemark, D.C. *Radio Sci.,* 1999, 34: 337-354.

Jacobson, A.R.; Cummins, K.L.; Carter, M.; Klingner, P.; Roussel-Dupré, D.; Knox, S.O. *J. Geophys. Res.*, 2000, 105, 15653-15662.

Jacobson, A.R.; Holzworth, R.H.; Harlin, J., Dowden, R.L.; Lay, E. H. *J. Atmos. Oceanic Tech., 2006,* 23, 1082-1092.

Janischewskyj, W.; Hussein, A. M.; Shostak, V.; Russan, I.; Li, J.-X.; Chang, J.-S. *IEEE Transactions on Power Delivery*, 1997, 12, no. 3.

Jerauld, J.; Rakov, V. A.; Uman, M. A.; Rambo, K. J.; Jordan, D. M.; Cummins, K. L.; Cramer, J. A. *J. Geophys. Res.*, 2005, 110, doi:10.1029/2005JD005924.

Kai-Hing, Y. *Importance Notes*, 2003, Hong Kong Observatory Report.

Kandalgaonkar, S.S.; Tinmaker, M.I.R.; Nath, A.; Kulkarni, M.K.; Trimbake, H.K. *Atmósfera*, 2005, 91-101.

Kasemir, H.W. *J. Geophys. Res.,* 1960, 65, 1873-1878.

Kilinc, M.; Beringer, J. *J. Climate,* 2007, 20, 1161-1173.

Kirkland, M. W.; Suszcynsky, D. M.; Guillen, J. L. L.; Green, J. L. *J. Geophys. Res.*, 2001, 106, 33499 –33509.

Koshak, W. J.; Solakiewicz, R.J.; Blakeslee, R.J. *J. Atmos. Ocean. Technol.*, 2004, 21, 543-558.

Kotaki, M.; Katoh, C. *J. Atmos. Terr. Phys.*, 1983, 45, 833–847.

Krehbiel, P.R.; Riousset, J.A.; Pasko, V.P.; Thomas, R.J.; Rison, W.; Stanley, M.A.; Eddens, H.E. *Nature Geoscience*, 2008, 1, 233-237.

Krider, E.P.; Noggle, R.C.; Uman, M.A. *J. Appl. Meteor.*, 1976, 15, 301-306.

Krider, E.P.; Noggle, R.C.; Pifer, A.E; Vance, D.L. Bull. *Amer. Meteor. Soc.*, 1980, 61, 980-986.

Kuleshov, Y.; Mackerras, D.; Darveniza, M. *J. Geophys. Res.*, 2006, 111, doi:10.1029/2005JD006982.

Lacerda, M.; Pinto Jr., O.; Pinto, I.R.C.A.; Diniz, J.H.; Carvalho, A.M. *Proceedings of 11th International Conference on Atmospheric Electricity (ICAE), 1999,* Huntsville, Alabama.

Latham, J.; Christian, H. Quart. J. Roy. *Met. Soc.,* 1998, 124, 1771-1773.

Lay, E. H.; Holzworth, R. H.; Rodger, C. J.; Thomas, J. N.; Pinto, O.; Dowden, R. L. *Geophys. Res. Lett.*, 2004, 31, 10.1029/2003GL018882.

Lee, S.C.; Lim, K.K.; Meiappa, M.; Liew, A.C. *IEEE Trans. Power App. Systems,* 1979, PAS-98, 1669-1674.

Lewis, E.A.; Harvey, R.B.; Rasmussen, J.B. *J. Geophys. Res.*, 1960, 63, 1879-1905.

Liu, X.; Zhang, Y. *Trans. IEE Japan*, 1998, 118-B, 170-175.

Loboda, M.; Betz, H.D.; Baranski, P.; Wiszniowski, J.; Dziewit, Z. *Proceedings of the 29th International Conference on Lightning Protection (ICLP), 2008*, Uppsala, Sweden.

Lyons, W.A.; Moon, D.A.; Schuh, J.A.; Pettit, N.J.; Eastman, J.R. *Proceedings of the Int. Conf. on Lightning and Static Electricity, 1989*, Bath, England.

Lyons, W.A.; Keen, C.S. *Mon. Wea. Rev.*, 122, 1994, 1897-1916.

Lyons, W.A.; Nelson, T.E.; Williams, E.R.; Cramer, J.A.; Turner, T.R. *Science*, 1998a, 282, 77-80.

Lyons, W.A.; Uliasz, M.; Nelson, T.E. *Mon. Wea. Rev.,* 1998b, 126, 2217-2233.

MacGorman, D.R.; Rust, W.D. *The Electrical Nature of Storms*, 1998, 422 p., Oxford University Press.

Mackerras, D. *Electr. Eng. Trans. Int. Eng.* Aust., 1978, EE14: 73-77.

Mackerras, D. *J. Geophys. Res.*, 1985, 90, 6195-6201.

Mackerras, D.; Darveniza, M. *J. Geophys. Res.*, 1994, 99, 10813-10821.

Malan, D.J. *Physics of lightning,* 1963, London, The English Universities Press.

Markson, R. Bull. *Amer. Meteor. Soc.*, 2007, 224, 233-241.

Markson, R.; Price, C. *Atmos. Res.*, 1999, 51, 309-314.

Marriot, W. Q. J. R. *Meteorol. Soc.,* 1908, 34, 210.

Marsh, P. T.; Brooks, H. E.; Karoly, D. *J. Atmos. Science Letters*, 2007, 8, doi: 10.1002/asl.159.

Martin, R.V.; Jacob, D.J.; Logan, J.A.; Bey, I.; Yantosca, R.M.; Staudt, A.C.; Li, Q.; Fiore, A.M.; Duncan, B.N.; Liu, H. *J. Geophys. Res.*, 2002, 107, doi:10.1029/2001JD001480.

Mazur, V. C. R. *Physique,* 2002, 3, 1393-1409.

Mazur, V.; Runke, L.H. *J. Geophys. Res.*, 1993, 94, 12,913-12,930.

Mazur, V.; Krehbiel, P.R.; Shao, X.M. *J. Geophys. Res.*, 1995, 100, 25,731-25,753.

McEachron, K.B. *J. Franklin Inst.*, 1939, 227, 149-217.

McEachron, K.B. *AIEE Trans.*, 1941, 60, 885-890.

McNutt, S.R.; Davis, C.M. *J. Volcanol. Geoth. Res.*, 2000, 102, 45-65.

Mello, J.C.D.; Schroeder, M.A.O.; Visacro Filho, S. *Proceedings of the Ground 2000 International Conference on Grounding and Earthling, 2000*, Belo Horizonte, Brazil.

Michalon, N.; Nassif, A.; Saouri, T.; Royer, J.F.; Pontikis, C.A. *Geophys. Res. Lett.*, 1999, 26, 3097-3100.

Mickley, L.J.; Jacob, D.J.; Rind, D. *J. Geophys. Res.*, 2001, 106, 3389-3399.

Ming, M.A.; Shanchang, T.A.O.; Baoyou, Z.H.U.; Weitao, L.U.; Yongbo, T.A.N. *Chinese Sci. Bull.*, 2005, 50, 2640-2644.

Miyake, K.; Suzuki, T.; Shinjou, K. *IEEE Trans. Pow. Del.*, 1992, 7, 1450-1456.

Molinari, J.; Knight, D.; Dickinson, M.; Vollaro, D.; Skubis, S. *Mon. Wea. Rev.*, 1994, 125, 2699-2708.

Molinari, J.; Moore, P.; Idone, V. *Mon. Wea. Rev.*, 1999, 127, 520–534.

Molinié, J.; Pontikis, C.A. *Geophys. Res. Lett.*, 1995, 22, 1085-1088.

Montandon, E. *Proceedings of the 21th International Conference on Lightning Protection (ICLP), 1992,* Berlin, Germany.

Murphy, M.J.; Holle, R.L. *Weather and forecasting*, 2005, 20, 125-133.

Murphy M.; Pifer A.; Cummins, K.; Pyle, R.; Cramer, J. *Proceedings of 17th International Lightning Detection Conference (ILDC)*, 2002, Tucson, AZ.

Murray, N.D.; Orville, R.E.; Huffines, G.R. *Geophys. Res. Lett.*, 27, 2000, 2249-2252.

Naccarato, K.P. PhD. *Thesis, INPE,* 2005, 258 p. (in Portuguese).

Naccarato, K.P.; Pinto Jr., O. *Atmos. Res.*, 2008, in press.

Naccarato, K.; Pinto Jr., O. *Proceedings of the International Lightning Detection Conference, 2008,* Tucson.

Naccarato, K.P.; O. Pinto Jr.; I.R.C.A. Pinto, *Geophys. Res. Lett.*, 2003, 30, 1674-1677.

Naccarato, K.P.; Pinto Jr., O.; Pinto, I.R.C.A. *Proceedings of the 1st International Conference on Lightning Physics and Effects (LPE), 2004*a, Belo Horizonte, Brazil.

Naccarato, K.P.; Pinto Jr., O.; Pinto, I.R.C.A. *Proceedings of the 18th International Lightning Detection Conference (ILDC), 2004*b, Vaisala, Helsinki, Finland.

Naccarato, K.P.; Pinto Jr., O.; Pinto, I.R.C.A. *Proceedings of the 19th International Lightning Detection Conference (ICLP), 2006*a, Vaisala, Tucson, AZ.

Naccarato, K.P.; Pinto Jr., O.; Pinto, I.R.C.A. *Proceedings of the 2nd International Conference on Lightning Physics and Effects (LPE), 2006*b, Maceió, Brazil.

Naccarato, K.P.; Pinto Jr., O.; Holzworth, R.H.; Blakeslee, R. *Proceedings of the 29th International Conference on Lightning Protection (ICLP), 2008,* Uppsala, Sweden.

Nag, A.; Jerauld, J.; Rakov, V.A.; Uman, M.A.; Rambo, K.J.; Jordam, D.M.; De Carlo, B.A.; Howard, J.; Biagi, C.J.; Hill, D.; Cummins, K.L.; Cramer, J.A. *Proceedings of the 29th International Conference on Lightning Protection, 2008*b, Uppsala, Sweden.

Nakamura, K.; Horii, K.; Kito, Y.; Wada, A.; Ikeda, G.; Sumi, S.; Yoda, M.; Aiba, S.; Sakurano, H. Wakamatsu, K. *IEEE Trans. Power Del,*. 1991, 6, 1311-1318.

Nakamura, K.; Horii, K.; Nakano, M.; Sumi, S. Res. *Lett. Atmos. Electr.*, 1992, 12, 29-35.

Newman, M.M. *Problems of Atmospheric and Space Electricity,* 1965, pp. 482-490, New York, Elsevier.

Ndlovu, N.; Evert, C.R. *Proceedings of the 19th International Lightning Detection Conference (ILDC), 2006*a, Tucson, AZ.

Nicholls, N. *Geophys. Res. Lett.*, 2008, 35, doi:10.1029/2008GL034499.

Nicholls, N.; Alexander, L. Progress Phys. Geography., 2007, 31, 77-87.

Nikolaenko, A.P.; Rabinowscz, L.M. *J. Atmos. Terr. Phys.*, 1995, 57, 1345-1348.

Ogawa, T. *Handbook of Atmospheric Electrodynamics,* 1995, vol. I, Ed. H. Volland, Boca Raton, FL, CRC Press pp. 23-63.

Orville, R. E. *J. Geophys. Res.,* 1991, 96, 17135–17142.

Orville, R.E. *Handbook of Atmospheric Electrodynamics,* 1995, vol. I, Ed. Hans Volland, Boca Raton, FL, CRC Press, pp. 137-149.

Orville R.E.; Spencer, D.W. Mon.Wea.Rev., 107, 934-943.

Orville R.E.; Henderson, R.W. *Mon. Wea. Rev.,* 1986, 114, 2640-2653.

Orville, R.E.; Huffines, G.R. *Mon. Wea. Rev.*, 2001, 129, 1179-1193.

Orville, R.E.; Zipser, E. J.; Brook, M.; Weidman, C.; Aulich, G.; Krider, E.P.; Christian, H.; Goodman, S.; Blakeslee, R.; Cummins, K. Bull. Am. Meteorol. Soc., 1997, 1055-1067.

Orville, R.E.; Huffines, G. R.; Burrows, W. R.; Holle, R. L.; Cummins, K. L. *Mon. Wea. Rev.,* 2002, 130, 2098-2109.

Petersen, W.A.; Rutledge, S.A. *J. Geophys. Res.*, 1992, 97, 11553-11560.

Petersen, W.A.; Buechler, D. *Proceedings of the American Meteorological Society Meeting, 2007*, New Orleans.

Petersen, W.A.; Christian, H.C.; Rutledge. S. A. *Geophys. Res. Lett.*, 2005, 32, doi:10.1029/2005GL023236.

Pinto, I.R.C.A.; Pinto Jr., O.; Rocha, R.M.L.; Diniz, J.H.; Carvalho, A.M. ; Cazetta Filho, A. *J. Geophys. Res.*, 1999a, 104, 24, 31381-31388.

Pinto, I.R.C.A.; Pinto Jr., O. *J. Atmos. Solar-Terr. Physics,* 2003, 65, 733-737.

Pinto, I.R.C.A.; Pinto Jr., O.; Gomes, M.A.S.S.; Ferreira, N.*J. Ann. Geophysicae,* 2004, 22, 697-700.

Pinto Jr., O. *Proceedings of VII International Symposium on Lightning Protection,* 2003, Curitiba, Brazil.

Pinto Jr., O. *The art of war against lightning,* 2005, São Paulo, Oficina de Texto, (in Portuguese).

Pinto Jr., O. *Proceedings of the International Lightning Detection Conference (ILDC),* 2008a, Tucson.

Pinto Jr., O. *Proceedings of the International Lightning Detection Conference (ILDC),* 2008b, Tucson.

Pinto Jr., O. *Atmospheric Science Research Progress,* Ed. C.-H. Yang, 2008c, Nova Publishers, in press.

Pinto Jr., O.; Pinto, I.R.C.A. J. Geophys. Res., 2008a, 113, doi:10.1029/2008JD009841.

Pinto Jr., O.; Pinto, I.R.C.A. *Proceedings of the 3rd International Conference on Lightning Physics and Effects, 2008*b, Florianopolis, Brazil.

Pinto Jr., O.; Gin, R.B.B.; Pinto, I.R.C.A.; Mendes Jr., O.; Diniz, J.H.; Carvalho, A.M. *J. Geophys. Res.*, 1996, 101, 29627-29635.

Pinto Jr., O.; Pinto, I.R.C.A.; Lacerda, M.; Carvalho, A.M.; Diniz, J.H.; Cherchiglia, L.C.L. *J. Atmos. Solar-Terr. Phys.*, 1997, 59, 1881-1883.

Pinto Jr., O.; Pinto, I.R.C.A.; Gin, R.B.B.; Mendes Jr., O. *Ann. Geophysicae,* 1998, 16, 353-355.

Pinto Jr., O.; Pinto, I.R.C.A.; Gomes, M.A.S.S.; Vitorello, I.; Padilha, A.L.; Diniz, J.H.; Carvalho, A.M.; ; Cazetta Filho, A. *J. Geophys. Res.*, 1999b, 104, 24, 31369-31380.

Pinto Jr. O.; Pinto, I.R.C.A.; Faria, H.H. *Geophys. Res. Lett.*, 2003a, 30, 1029-1032.

Pinto Jr., O.; Pinto, I.R.C.A.; Diniz, J.H.; Filho, A.C.; Carvalho, A.M.; Cherchiglia, L.C.L. J. *Atmos. Terr. Phys.*, 2003b, 65, 739-748.

Pinto Jr., O.; Naccarato, K.P.; Pinto, I.R.C.A.; Saba, M.M.F.; Gardiman, V.L.G.; de M. Garcia, S.A.; Abdo, R. F.; Assunção, L.A.R. *Proceedings of the International Conference on Lightning Physics and Effects (LPE), 2004*a, Belo Horizonte, Brazil.

Pinto Jr., O.; Saba, M.M.F.; Pinto, I.R.C.A.; Tavares, F.S.S.; Naccarato, K.P.; Solorzano, N.N.; Taylor, M.J.; Pautet, P.D.; Holzworth, R.H. *Geophys. Res. Lett.*, 2004b, 31, doi:10.1029. 2004GL020264.

Pinto Jr., O.; Pinto, I.R.C.A.; Saba, M.M.F.; Solorzano, N.N.; Guedes, D. *Atmos. Res.*, 2005, 76, 493-502.

Pinto Jr., O.; Naccarato, K.P.; Saba, M.M.F.; Pinto, I.R.C.A.; Abdo, R. F.; de M. Garcia, S.A.; Cazetta Filho, A. *Proceedings of the 19th International Lightning Detection Conference (ILDC), 2006a*, Tucson, AZ.

Pinto Jr., O.; Naccarato, K. P.; Pinto, I. R. C. A.; Fernandes, W. A.; Neto, O. P. *Geophys. Res. Lett.*, 2006b, 33, doi:10.1029/2006GL026081.

Pinto Jr., O.; Pinto, I.R.C.A.; Naccarato, K.P. *Atmos. Res.*, 2007, 84, 189-200.

Pinto Jr., O.; Pinto, I.R.C.A.; Saba, M.M.F.; Naccarato, K.P. Lightning: *Principles, Instruments and Applications*, 2008, Ed., H.-D Betz, U. Schumann and P. Laroche, Springer, in press.

Prentice, S.A. *Lightning, 1977*, vol. 1, Physics of Lightning, ed. R. H. Golde, New York, Academic Press, pp. 465-495.

Price, C. *Geophys. Res. Lett.*, 1993, 20, 1363-1366.

Price, C. *Proceedings of the 29th International Conference on Lightning Protection (ICLP), 2008*, Uppsala, Sweden.

Price, C.; Rind, D. *J. Geophys. Res.*, 1992, 97, 9919-9933.

Price, C.; Rind, D. Mon, *Wea. Rev.*, 1994, 122, 1930-1939.

Price, C.; Melnikov, A. J. *Atmos. Solar-Terr. Phys.*, 2004, 66, 1179-1185.

Price, C.; Asfur, M. Bull. *Amer. Meteor. Soc.*, 2006a, 87, 291-298.

Price, C.; Asfur, M. *Earth Planets Space,* 2006b, 58, 1-5.

Price, C.; Penner, J.; Prather, M. *J. Geophys. Res.,* 1997, 102, 5929-5941.

Rachidi, F.; Bermudez, J.L.; Rubinstein, M.; Rakov, V.A. *J. Electrostatics,* 2004, 60, 121-129.

Raizman, S.; Mendez, Y.; Vivas, J.; Arévalo, J. *Proceedings of IV Congress of Electrical Engineer, 2004*, Caracas (in Spanish).

Rajeevan, M.; Bhate, J.; Jaswal, A. K., Geophys. Res. Lett., 2008, 35, doi:10.1029/ 2008GL035143.

Rakov, V.A. *Recent Res. Devel. Geophysics,* 1999, 2, 141-171.

Rakov, V.A. *IEEE Transactions on Electromagnetic Compatibility*, 2001, 43, 654-661.

Rakov, V.A. *Recent Res. Devel. Geophysics*, 2003, 5, 57-71.

Rakov, V.A. *Recent Res. Devel. Geophysics*, 2004, 6, 1-35.

Rakov, V.A. *Proceedings of the Symposium on Lightning Protection (SIPDA), 2007*, Foz do Iguaçu, Brazil.

Rakov, V. A.; Uman, M. A. *J. Geophys. Res.*, 1990, 95, 5455-5470.

Rakov, V. A.; Uman, M. A. *Lightning: Physics and Effects, 2003*, Cambridge University Press, Cambridge, 687p.

Rakov, V. A.; Huffines, G. R. *J. Applied Meteorology,* 2003, 42, 1455-1462.

Rakov, V.A.; Uman, M.A.; Thottappillil, R., *J. Geophys. Res.*, 1994, 99, 10745-10750.

Rakov, V.A.; Uman, M.A.; Rambo, K.J.; Fernandez, M.I.; Fisher, R.J.; Schnetzer, G.H.; Thottappillil, R.; Eybert-Berard, A.; Berlandis, J.P.; Lalande, P.; Bonamy, A.; Laroche, P.; Bondiou-Clergerie, A. *J. Geophys. Res.*, 1998, 103, 14117-14130.

Reeve, N.; Toumi, R. Quart. *J. Royal Met. Soc.*, 1999, 125, 893-903.

Risk, F.A.M., *IEEE Trans. Power Delivery*, 1994, 9, 162-193.

Rison, W.; Thomas, R.J.; Krehbiel, P.R.; Hamlin, T.; Harlin, *J. Geophys. Res. Lett.*, 1999, 26, 3573-3576.

Rison, W.; Hamlin, T.; Harlin, *J. Atmos. Res.*, 2005, 76, 247-271.

Rodger, C.J.; Brundell, J.B.; Dowden, R.L.; Thomson, N.R. *Ann. Geophys.*, 2004, 22, 747-758.

Rodger, C.J.; Brundell, J.B.; Dowden, R.L.; Thomson, N.R. *Ann. Geophys.*, 2005, 23, 277-290.

Rodriguez, C.A.M.; Sales, F.; Câmara, K. S.; Pinheiro, F. G.; Anagnostou, E. E. *Proceedings of the Symposium on Lightning Protection (SIPDA), 2007,* Foz do Iguaçu, Brazil.

Rompala, J.T.; Blakeslee, R.J.; Bailey, J.C. Proceedings of the 12th International Conference on Atmospheric Electricity (ICAE), 2003, Versailles, France.

Romualdo, C.; Brito, F.; Perez, H.; de la Rosa, F.; Sarmiento, H.G. *IEEE Comp. Appl. Power,* 1989, 2, 43-47.

Rubinstein, M. *Proceedings of the EMC Conference, 1995,* Zürich, Switzerland.

Rust, R.M.; MacGorman, D.R.; Arnold, R.T. *Geophys. Res. Lett.*, 1981, 8, 791-794.

Rust, W.D.; MacGorman, D.R.; Bruning, E.C.; Weiss, S.A.; Krehbiel, P.R.; Thomas, R.J.; Hamlin, T.; Harlin, *J. Atmos. Res.*, 2005, 76, 247-271.

Ryu, J.H.; Jenkins, G.S. *Atmos. Environment,* 2005, 39, 5799-5805.

Saba, M. M. F.; Pinto Jr., O.; Ballarotti, M. G.; Naccarato, K. P.; Cabral, G. F. *Proceedings of the International Lightning Detection Conference (ILDC), 2004,* Helsinki, Finland.

Saba, M.M.F.; Pinto Jr., O.; Solorzano, N.N.; Eybert-berard, A. *Atmos. Re.*, 2005, 76, 402-411.

Saba, M.M.F.; Ballarotti, M.G.; Pinto Jr., O. *J. Geophys. Res.*, 2006a, doi:10.1029/2005JD006415.

Saba, M.M.F.; Pinto Jr., O.; Ballarotti, M.G. *Geophys Res Lett.*, 2006b, doi:10.1029/2006GL027455.

Saba, M.M.F.; Ballarotti, M. G.; Campos, L. Z. S.; Pinto Jr., O. *Proceedings of the International Symposium on Lightning Protection (SIPDA), 2007a,* Foz do Iguaçu, Brazil.

Saba, M.M.F.; Ballarotti, M.G.; Pinto Jr., O.; Ferro, M. A. S. *Proceedings of the International Conference on Atmospheric Electricity (ICAE), 2007b,* Beijing, China.

Saba, M. M. F.; Campos, L.Z. S.; Ballarotti, M.; Pinto Jr., O. *Proceedings of the International Conference on Atmospheric Electricity (ICAE), 2007c,* Beijing, China.

Saba, M. M. F.; Cummins, K.L.; Warner, T. A.; Krider, E. P.; Campos, L. Z. S.; Ballarotti, M. G.; Pinto Jr., O.; Fleenor, S. A. *Proceedings of the International Lightning Detection Conference, 2008a,* Tucson, AZ.

Saba, M.M.F.; Cummins, K.L.; Warner, T.A.; Krider, E.P.; Campos, L.Z.S.; Ballarotti, M.G.; Pinto Jr., O.; Fleenor, S.A. *Geophys Res Lett.*, 2008b, doi:10.1029/2007GL033000.

Saba, M.M.F.; Campos, L. Z. S.; Ballarotti, M.G.; Pinto Jr., O. *Proceedings of the International Conference on Lightning Protection, 2008c*, Uppsala, Sweden.

Sanderson, M. G.; Collins, W. J.; Johnson, C. E.; Derwent, R. G. *Atmos. Environ.*, 2006, 40, 1275-1283.

Saraiva, A. C. V.; Saba, M. M. F.; Campos, L. Z. S.; Pinto Jr., O.; Cummins, K. L.; Krider, E. P.; Fleenor S. A. *Proceedings of the International Lightning Detection Conference, 2008a*,Tucson, AZ.

Saraiva, A. C. V.; Saba, M. M. F.; Pinto Jr., O.; Cummins, K. L.; Krider, E. P.; Campos, L. Z. S. *Proceedings of the III International Conference on Lightning Physics and Effects, 2008b*, Florianópolis, Brazil.

Sato, M.; Fukunishi, H. *Geophys. Res. Lett.*, 2003, doi:10.1029/2003GL017291.

Sátori, G.; Williams, E.; Lemperger, I. *Proceedings of the International Conference on Atmospheric Electricity (ICAE), 2007,* Beijing, China.

Sátori, G.; Williams, E.; Lemperger, I. Atmos. Res., 2008, in press.

Schonland, B.F.J. *Encyclopedia of Physics*, 1956, vol. 22, Berlin, Springer-Verlag, pp. 576-628.

Schroeder, M.A.O.; Soares Jr., A.; Visacro Filho, S.; Cherchiglia, L. C. L.; Souza; V. J.; Diniz, J.H.; Carvalho, A.M. *Proceedings of the V International Symposium on Lightning Protection (SIPDA), 1999*, São Paulo, Brazil.

Schulz, W. PhD Thesis, 1997, *Technical University of Vienna, Faculty of Electrical Engineering, Austria.*

Schulz, W.; Diendorfer, G. *Proceedings of the 15th International Lightning Detection Conference (ILDC), 1998,* Tucson, AZ.

Schulz, W.; Diendorfer, D. *Proceedings of the International Lightning Detection Conference (ILDC), 2006*, Tucson, AZ.

Schulz, W.; Cummins, K.; Diendorfer, G.; Dorninger, M. *J. Geophys. Res.,* 2005, 110, doi:10.1029/2004JD005332.

Schulz, W.; Sindelar, S.; Kafri, A.; Gotschl, T.; Theethayi, N.; Thottappillil, R. *Proceedings of the 29th International Conference on Lightning Protection (ICLP), 2008,* Uppsala, Sweden.

Schumann, W.O. *Z. Naturforsch*, 1952, 7a, 149-152.

Schumann, U.; Huntrieser, H. *Atmos. Chem. Phys.*, 2007, 7, 3823-3907.

Sekiguchi, M.; Hayakawa, M.; Nickolaenko, A.P.; Hobara, Y. *Ann. Geophys.*, 2006, 24, 1809-1817.

Sentman, D. D. *Handbook of Atmospheric Electrodynamics, 1995*, Ed. H. Volland, p. 267, CRC Press, Boca Raton, FR.

Sentman, D. D.; Fraser, B. J. *J. Geophys. Res.*, 1991, 96, 15973-15978.

Shao, X.M.; Krehbiel, P.R.; Thomas, R.J.; Rison, W., *J. Geophys. Res.*, 1995, 100, 2749-2783.

Shao, X.M.; Harlin, J.; Stock, M.; Stanley, M.; Regan, A.; Wiens, K.; Hamlin, T.; Pongratz, M.; Suszcynsky, D.; Light, T. *EOS Trans.*, 2005, 86, 398.

Sharp, A.J. *Proceedings of the International Conference on Atmospheric Electricity (ICAE), 1999,* Hunstville, AL.

Sherwood, S. C.; Phillips, V. T. J.; Wettlaufer, J. S. *Geophys. Res. Lett.*, 2006, 33, doi:10.1029/2005GL025242.

Shindell, D.T.; Faluvegi, G.; Unger, N.; Aguilar, E.; Schimidt, G.A.; Koch, D.M.; Bauer, S.E.; Miller, R.L. *Atmos. Chem. Phys.*, 2006, 6, 4427-4459.

Shindo, T.; Yokoyama, S. IEEE Trans. Power Delivery, 1998, 13, 1368-1474.

Shindo T.; Uman, M. A. *J. Geophys. Res.*, 1989, 94, 5189-5198.

Singh, H.B.; Thompson, A.M.; Schlager, H. *Geophys. Res. Lett.*, 1999, 26, 3053-3056.

Smith, D.A.; Shao, X.M.; Holden, D.N.; Rhodes, C.T.; Brook, M.; Krehbiel, P.R.; Stanley, M.; Rison, W.; Thomas, R.J. *J. Geophys. Res.*, 1999, 104, 4189-4212.

Soetjipto, S.; Iskanto, E.; Soekarto, *J. Proceedings of the 6th ISH, 1989*, New Orleans.

Soetjipto, S.; Sunoto, M.A.; Iskanto, E.; Hasan, M.; Hendi, W. *Proceedings of the 8th ISH, 1993*, Yokohama, Japan.

Solomon, R.; Baker, M. Mon. *Wea. Rev.*, 1994, 122, 1878-1886.

Solorzano, N.N. PhD Thesis, 2003, INPE, 178 p. (in Portuguese).

Soriano, L.R.; de Pablo; F.; Diez, E.G. *J. Geophys. Res.*, 2001, 106, 11891-11901.

Soriano, L.R.; de Pablo, F.; Tomas, C. *J. Atmos. Terr. Phys.*, 2005, doi:10.1026/j.jastp.2005.08.019.

Souza, P.E.; Pinto Jr., O.; Pinto, I.R.CA.; Ferreira, N. J.; dos Santos, A. F. *Atmos. Res.*, 2008, in press.

Steiger, S.M.; Orville, R.E.; Huffines, G. *J. Geophys. Res.*, 2002, doi:10.1029/2001JD001142.

Stevenson, D. S.; Doherty, R.; Sanderson, M.; Johnson, C. E.; Collins, B.; Derwent, R. G. *Faraday Discuss.*, 2005, doi:10.1039/b417412g.

Stevenson, D.; Dentener, F. J.; Schultz, M. G.; Ellingsen, K.; Van Noije, T. P. C.; Wild, O.; Zeng, G.; Amman, M.; Atherton, C. S.; Bell, N.; Bergmann, D. J.; Bey, I.; Butler, T.; Cofala, J.; Collins, W. J.; Derwent, R. G.; Doherthy, R. M.; Drevet, J.; Heskes, H. J.; Fiore, A. M.; Gauss, M.; Hauglustaine, D. A.; Horowitz, L. W.; Isaksen, I. S. A.; Krol, M. C.; Lamarque, J.-F.; Lawrence, M. G.; Montanaro, V.; Muller, J.-F.; Pitari, G.; Prather, M. J.; Pyle, J. A.; Rast, S.; Rodriguez, J. M.; Sanderson, M. G.; Savage, N. H.; Shindell, D. T.; Strahan, S. E.; Sudo, K.; Szopa, S. *J. Geophys. Res.*, 2006, doi:10.1029/2005JD006338.

Stocks, B.J.; Fosberg, M. A.; Lynham, T. J.; Mearns, L.; Wotton, B. M.; Yang, Q.; Jin, J-Z.; Lawrence, K.; Hartley, G. R.; Mason, J. A.; McKenney, D. W. *Clim. Change*, 1998, 38, 1–13.

Stolzenburg, M.; Rust, W.D.; Marshall, T.C. *J. Geophys. Res.*, 1998, 103, 14,079-14,096.

Suda, T.; Shindo, T.; Yokohama, S.; Tomita, S.; Wada, A.; Tanimura, A.; Honma, N.; Tanigushi, S.; Shimizu, M.; Sakai, T.; Sonoi, Y.; Yamada, K.; Komori, M.; Ikesue, K.; Toda, K. *Proceedings of the 26th International Conference on Lightning Protection, 2002*, Cracow, Poland.

Sunoto, A.M. *Proceedings of the 18th International Conference on Lightning Protection (ICLP)*, 1985, Munich, Germany.

Suszcynsky, D.M.; Kirkland, M.W.; Jacobson, A.R.; Franz, R.C.; Knox, S.O.; Guillen, J.L.L.; Green, J.L. *J. Geophys. Res.*, 2000, 105, 2191-201.

Tarazona, J.; Ferro; C.; Urdaneta, A.J. *Proceedings of the CIGRE Meeting, 2006*, Paris, France.

Théry, C., *Atmos. Res.*, 2001, 56, 397-409.

Thomas, R.J.; Krehbiel, P.R.; Rison, W.; Hamlin, T.; Boccippio, D.J.; Goodman, S.J.; Christian, H.J. *Geophys. Res. Lett.*, 2000, 27, 1703-1706.

Thompson, A. M.; Doddridge, B. G.; Witte, J. C.; Hudson, R. D.; Luke, W. T.; Johnson, J. E.; Johnson, B. J.; Oltmans, S. J.; Weller, R. *Geophys. Res. Lett.*, 2000, 27, 3317–3320.

Thomson, E.M.; Gali, M.A.; Uman, M.A.; Beasley, M.J. *J. Geophys. Res.*, 1984, 89, 4910-4916.

Toracinta, E.R.; Zipser, E. J. *J. Appl. Meteorol.*, 2001, 40, 983-1002.

Toracinta, E.R.; Cecil, D.J.; Zipser, E.J.; Nesbitt, S.W. Mon. Wea. Rev., 2002, 130, 802-824.

Torres, H.; Rondon, D.; Briceño, W.; Barreto, L. *Proceedings 23th International Conference on Lightning Protection (ICLP)*, 1966, Florence, Italy.

Torres, H.; Gallego, L.; Salgado, M.; Younes, C.; Herrera, J.; Quintana, C.; Rondón, D.; Pérez, E.; Montaña, J.; Vargas, M. *Proceedings of the International Symposium on Lightning Protection (SIPDA)*, 2001, Santos, Brazil.

Torres, H.; Williams, E.; Rakov, V.; Pinto Jr., O., Visacro, S.; Perez, E.; Younes, C.; Blanco, C.; Vasquez, C. *CIGRE report*, 2008, submitted.

Toumi, R.; Haigh, J. D.; Law, K. S. *Geophys. Res. Lett.*, 1996, 23, 1037-1040.

Trapp, R.J.; Diffenbaugh, N. S.; Brooks, H. E.; Baldwin, M. E.; Robinson, E. D.; Pal, J. S. Proceeding of the, National Academy of Sciences, 2007, doi: 10.1073/pnas.0705494104.

Turman, B. N; Edgar, B.C. *J. Geophys. Res.*, 1982, 87, 1191–1206.

Triginelli, W.A.C.; Carvalho, A.M.; Diniz, J.H.; Cherchiglia, L.C.L. Proceedings of the III *International Symposium on Lightning Protection*, 1995, São Paulo, Brazil.

Uman, M.A. *The art and science of lightning protection*, 2008, Cambridge University Press, 240 p.

Uman, M.A.; McLain, D.K.; Krider, E.P. *Am. J. Phys.*, 1975, 43, 33-38.

Ushio, T.; Heckman, S. J.; Boccippio, D. J.; Christian, H. J.; Kawasaki, Z. *J. Geophys. Res.*, 2001, 106, 24089-24095.

Vecchi, G.A.; Clement, A.; Soden, B.J. EOS Trans., 2008, 89, 81-83.

Velasco, I.; Fritch, J.M. *J. Geophys. Res.*, 1987, 92, 9591-9613.

Visacro, S.; Soares Jr., A.; Schoroeder, M.A.O.; Cherchiglia, L.C.L.; de Sousa, V.J. *J Geophys Res.*, 2004, 109, doi:10.1029/2003JD003662.

Visacro S.; F. H. Silveira, *Geophys. Res. Lett.*, 2005, doi:10.1029/2005GL023255.

Webster, P. J.; Holland, G. J.; Curry, J. A.; Chang, H.-R. *Science*, 2005, 309, 1844-1846.

Williams, E.R. *J. Geophys. Res.*, 1989, 94, 13,151-13,167.

Williams, E.R. *Science*, 1992, 256, 1184-1187.

Williams, E.R. Mon. *Wea. Rev.*, 1994, 122, 1917-1929.

Williams, E.R. Global Atmos. Ionos. Electromag. Phenomena associated with Earthquakes, 1999, Ed. M. Hayakawa, *TERRAPUB*, Tokyo, pp. 939-949.

Williams, E.R. *Atmos. Res.*, 2005, 76, 272-287.

Williams, E.R. *Plasma Sources Sci. Technol.*, 2006, 15, 91-108.

Williams, E.R. *Proceedings of the 13th International Conference on Atmospheric Electricity* (ICAE), 2007, Beijing, China.

Williams, E.R. *Atmos. Res.*, 2008, in press.

Williams, E.; Renno, N. Mon. *Wea. Rev.*, 1993, 121, 21-36.

Williams, E.R.; Stanfill, S. C.R. *Acad. Sci. Phys.*, 2002, 3, 1277-1292.

Williams, E.R.; Sátori, G. *J. Atmos. Solar-Terr. Phys.*, 2004, 66, 1213-1231.

Williams, E.R.; Rutledge, S.A.; Geotis, S.G.; Renno, N.; Rasmussen, E.; Rickenbach, A. *J. Atmos. Sci.*, 1992, 49, 1386-1395.

Williams, E.R.; Boldi, B.; Matlin, A.; Weber, M.; Hodanish, S.; Sharp, D.; Goodman, S.;
    Raghavan, R.; Buechler, D. *Atmos. Res.*, 1999, 51, 245-265.

Williams, E.R.; Patel, A.C.; Boldi, R. *EOS Trans.*, 2001, 82, F141.

Williams, E.R.; Rosenfeld, D.; Madden, N.; Gerlach, J.; Gears, N.; Atkinson, L.; Dunnemann,
    N.; Frostrom, G.; Antonio, M.; Biazon, B.; Camargo, R.; Franca, H.; Gomes, A.; Lima,
    M.; Machado, R.; Manhaes, S.; Nachtigall, L.; Piva, H.; Quintiliano, W.; Machado, L.;
    Artaxo, P.; Roberts, G.; Renno, N.; Blakeslee, R.; Bailey, J.; Boccippio, D.; Betts, A.;
    Wolff, D.; Roy, B.; Halverson, J.; Rickenbach, T.; Fuentes, J.; Avelino, *E. J. Geophys.
    Res.*, 2002, doi:10.1029/2001JD000380.

Williams, E.R.; Chan, T.; Boccippio, D. Islands as miniature continents: another look at the
    land-ocean lightning contrast, J. Geophys. Res., 2004, doi:10.1029/2003JD003833.

WMO (World Meteorological Organization) Publication, World Distribution of
    Thunderstorm Days, Part 2: *Tables of Marine Data and World Maps*, 1956, Report
    WMO/OMM-No. 21.

Yahaya, M.P.; Ahmad, H.B.; Alam, M.A. *Proceedings of National technical Seminar on
    Standardization and Development of Lightning Protection Technologies, 1996*, Petaling
    Jaya, Malaysia.

Ye, B.; Del Genio, A.D.; Lo, K.K.-*W. J. Climate*, 1998, 11, 1997-2015.

Younes, C.; Torres, H. ; Pérez, E.; Herrera, J.; Montaña, J.; Vargas, M.; Gallego, L.; Rondón,
    D.; Pavas, A.; Cajamarca, G.; Urrutia, D. *Proceedings of the International Symposium on
    Lightning Protection (SIPDA)*, 2003, Curitiba, Brazil.

Younes, C.; Torres, H.; Pérez, E.; Gallego, L.; Cajamarca, G.; Pavas, A. Proceedings of the
    International Conference on Lightning Protection (ICLP), 2004, Avignon, France.

Zajac, B.A.; Rutledge, S.A. *Mon. Wea. Rev.*, 2001, 129, 10,833-10,841.

Zhang, Y.; Li, J.; Lu, W.; Dong, W.; Zheng, D.; Chen, S.; Qiu, S.; Wang, T.; Liu, H.; Lan, Y.
    Proceedings of the 29th International Conference on Lightning Protection, 2008,
    Uppsala, Sweden.

Zipser, E.J. Mon. *Wea. Rev.*, 1994, 122, 1837-1851.

Zipser, E.J.; Cecil, D.J.; Liu, C.; Nesbitt, S.W.; Yorty, D.P. Bull. *Amer. Meteorol. Soc.*, 2006,
    87, 1057-1071.

INDEX